Azure Containers Explained

Leverage Azure container technologies for effective application migration and deployment

Wesley Haakman

Richard Hooper

BIRMINGHAM—MUMBAI

Azure Containers Explained

Group Product Manager: Rahul Nair

Publishing Product Manager: Surbhi Suman

Senior Editor: Tanya D'cruz

Technical Editor: Shruthi Shetty

Copy Editor: Safis Editing

Project Coordinator: Shagun Saini and Prajakta Naik

Proofreader: Safis Editing

Indexer: Tejal Daruwale Soni

Production Designer: Aparna Bhagat

Marketing Coordinator: Nimisha Dua

First published: February 2023

Production reference: 1030223

Published by Packt Publishing Ltd.
Livery Place
35 Livery Street
Birmingham
B3 2PB, UK.

ISBN 978-1-80323-105-1

`www.packtpub.com`

To my wife Amy and daughter Charlotte who continuously support my endeavors. You help me focus and make work/life balance a reality

– Wesley Haakman

To my wife Laura, without all the love and support of whom I would not be where I am today, and my three cats Oreo, Luna, and Toast, who kept me company during the writing of this book.

– Richard Hooper

Contributors

About the authors

Wesley Haakman lives in the Netherlands and has worked with Microsoft Azure for over 8 years. He is a Microsoft Azure MVP and works as the head of DevOps at Intercept, a cloud service provider company in the Netherlands. Wesley has worked in IT for more than 18 years, starting his career as an IT support technician for Novell NetWare and SUSE Linux environments. Now, he primarily focuses on using Microsoft Azure for building and helping DevOps teams to add value for their customers. In his spare time, Wesley enjoys writing his own blog, as well as providing sessions for events and user groups. Outside of tech, he loves mountain biking and spends a fair amount of time with his family.

Richard Hooper, also known as Pixel Robots online, lives in Newcastle, England. He is a Microsoft MVP and an architect for Azure at Intercept, which is based in the Netherlands. He has about 20 years of professional experience in the IT industry. He has worked with Microsoft technologies for his whole career, but he has also dabbled in Linux. Richard has a passion for learning about new technologies – you can check that out on his blog, *Pixel Robots*. He is very enthusiastic about the cloud-native space on Azure. In his spare time, he enjoys sharing knowledge and helping people with whatever technology he has hands-on experience with and is passionate about via his blog posts, podcasts, videos, and more.

About the reviewer

Suraj S. Pujari is a cloud solution architect at Microsoft India with more than 12 years of experience in IT. His technical capabilities comprise helping customers with their digital transformation, migration, solution design, and modernization on the cloud. He works with a wide range of small, medium-sized, and large businesses, spanning from banking to manufacturing. In his free time, he likes to do yoga and play with his little one.

Table of Contents

3

Deploying Containers to Azure Functions 31

4

Azure Container Instances for Serverless Containers 47

5

Azure Container Apps for Serverless Kubernetes 61

6

Part 2: Choosing and Applying the Right Technology

7

8

9

Container Technologies for Startups 125

10

Container Technologies for Small and Medium-Sized Businesses 139

11

Container Technologies for Enterprises 165

Part 3: Migrating Between Technologies and Beyond

Preface

Whether for a start-up or an enterprise, decisions related to using different container technology on Azure have an impact on your app migration and modernization strategies. The challenge that companies face is how to choose the right one and when to decide to move on to the next technology. *Azure Containers Explained* will help you to make the right architectural choices for your solutions and give you a deeper understanding of the migration path to other platforms using practical examples.

This book will begin with a recap on containers as a technology and where you can store them within Azure. You'll learn about the different container technologies offered by Microsoft Azure and understand how each platform – namely, Azure Container Apps, **Azure Kubernetes Service** (**AKS**), Azure Container Instances, Azure Functions, and Azure App Services – works. Each of them has its own characteristics and use case when it comes to implementation. You'll gain skills to build upon your own application of a container solution on Azure using the best practices from real-world examples and successfully transform your business from the start-up phase to the enterprise phase.

By the end of this book, you'll be able to effectively cater to your business and application needs by selecting and modernizing your applications using various Microsoft Azure container services.

Who this book is for

This book is intended for cloud and DevOps architects, application developers, as well as technical leaders, decision-makers, and IT professionals working with Microsoft Azure and cloud-native technologies, especially containers. Reasonable knowledge of containers and a good understanding of Microsoft Azure will be helpful when using this book.

What this book covers

Chapter 1, Azure Containers – Pleased to Meet You, explains how enormous the container landscape on Azure is. Finding your way through it can be challenging. What container technologies are there, which ones can you leverage, and what tools do you need? In this chapter, we will discuss the technologies Microsoft Azure provides when it comes to container technologies, the popular and common use cases for them, and the tools you need to get started.

Chapter 2, Azure App Services – Running a Container Was Never That Difficult, highlights the capabilities of Azure App Services to run containerized solutions. We will discuss the pros and cons and use cases for it and explain how someone would get started with hosting their first container on Azure

App Services. Using the samples provided (GitHub-hosted), we will explain how to run a stable and reliable configuration on Microsoft Azure.

Chapter 3, Deploying Containers to Azure Functions, highlights the capabilities of Azure Functions to run containerized solutions. As in the previous chapter, we will discuss the pros and cons, the use cases for it, and get started with hosting your first container on Azure Functions. Finally, we will discuss how to run a stable and reliable configuration on Microsoft Azure.

Chapter 4, Azure Container Instances for Serverless Containers, discusses the capabilities of Azure Container Instances to run containerized solutions. Similar to the previous chapters, we will cover the advantages and disadvantages, the use cases and then get started with hosting their first container on Azure Container Instances. You will also learn how to run a reliable configuration on Microsoft Azure.

Chapter 5, Azure Container Apps for Serverless Kubernetes, covers the capabilities of Azure Container Apps to run containerized solutions. We will discuss its pros, cons, and a few use cases, following which we will get started with hosting your first container on Azure Container Apps. Using the samples provided on GitHub, we will explain how to run a stable configuration on Microsoft Azure.

Chapter 6, Azure Kubernetes Service for Kubernetes in the Cloud, highlights the capabilities of Azure Kubernetes Service to run containerized solutions. Like in the previous chapters, we will discuss the pros and cons and the use cases for it and explain how someone would get started with hosting their first container on Azure Kubernetes Service. We will also explain how to run a stable and reliable configuration on Microsoft Azure using the samples provided on GitHub.

Chapter 7, The Single Container Use Case, will dig into the details. Let's say you have a solution that runs on a single container (a single image). Whether it is because you only require a single container or because you are just getting started in the world of containers, you will need to deploy and run this thing. Maybe you're even coming from a different technology and you want to modernize it using container technologies. In this chapter, we will dive into what that requires, which of the technologies discussed in *Part 1* will fit your requirements, and how you can leverage them successfully.

Chapter 8, Deciding the Best-Fitting Azure Technologies for Multiple Containers, will then discuss what you can do when your requirements call for more than just one container. Maybe your solution consists of multiple containers. Different technologies on Microsoft Azure support different configurations. In this chapter, we will explain multiple scenarios that require multiple containers to run and which Microsoft Azure technologies are the best fit for these scenarios.

Chapter 9, Container Technologies for Startups, takes you a step further. You've experimented with the technologies, so you know whether you want to run a single-container solution or a multiple-container solution. Let's map this to the real world. You're a start-up and you want to run your solution on Microsoft Azure. As a start-up, your company has particular requirements in terms of costs, ease of use, scalability, and availability. Which technologies will be the best fit for you as a start-up? How can you leverage these technologies on Microsoft Azure to quickly get started building your start-up?

Chapter 10, Container Technologies for Small and Medium-Sized Businesses, goes even further than the previous chapter. You've outgrown the start-up phase, you have customers with certain requirements, you have your own business requirements, and, let's be clear, your business has matured. Every new era involves a degree of upheaval. During this chapter, we will dive into the different solutions for small and medium-sized businesses, what it means to move from a "start-up solution" to a more advanced one, and which technologies will be the best fit for small and medium-sized businesses. Going from being a start-up to a small or medium-sized business means that you need to reconsider the decisions you made before, which isn't such a bad thing.

Chapter 11, Container Technologies for Enterprises, will discuss a scenario in which you've become really successful, you're growing fast, and your business is starting to look like an enterprise. That means new requirements, which might require different technologies. The big difference between start-ups and enterprises is having stricter requirements from customers in the latter case, but you also have more financial momentum. During this chapter, we will look into these new requirements and what that means for the container technologies you can use on Azure. We'll think about compliance, security, scalability, and minimizing overhead.

Chapter 12, Migrating Between Container Technologies, applies to when you have deployed your container solution on an Azure technology but it doesn't fit the use case anymore. Businesses change and therefore requirements change. In this chapter, we will recap what we have implemented based on our use cases so far and summarize the technical constraints we have introduced while doing so.

Chapter 13, Azure Container Instances – I Like the Scalability, But I Need More, explores what you can do when you're using Azure Container Instances to run your container(s). As we established in *Part 2* of this book, requirements will change over time. What will the next step be when Azure Container Instances no longer fits your requirements, and what will the migration path to different technologies look like?

Chapter 14, Beyond Azure Container Apps, builds on your need to use Azure Container Apps to run your container(s). We now know for certain that requirements will change over time. What will the next step be when Azure Container Apps no longer fits your requirements and what will that migration path to different technologies look like?

Chapter 15, Azure Kubernetes Service – The Next Steps, will take things to the next level. You now use AKS to run your container(s). What will the next step be? You have already migrated to AKS, so is there anything else? Of course, there is! Once you have successfully deployed to AKS, the wider world of integrating with other Azure and open source technologies has opened up! There are many things to explore here, which we will discuss in this chapter. We'll discuss the cloud-native landscape, how you keep track of everything that approaches, and how you go about fitting that into your business.

Chapter 16, What's Next – Diving into Future Trends and More, will look at what we consider the future trends for containers running on Microsoft Azure. We will provide you with information on where you can go to keep up to date with what's coming and where you can go to learn more about each technology and even get certifications for using them.

To get the most out of this book

You will need an Azure account with Owner permissions on the subscription and a global admin on the tenant.

Software/hardware covered in the book	Operating system requirements
Azure Containers	Windows, macOS, or Linux
Docker	N/A
Kubernetes	N/A

After you have finished this book, the authors would love to hear from you, so please do reach out and discuss containers on Azure with us.

Download the example code files

You can download the example code files for this book from GitHub at `https://github.com/PacktPublishing/Azure-Containers-Explained`. If there's an update to the code, it will be updated in the GitHub repository.

We also have other code bundles from our rich catalog of books and videos available at `https://github.com/PacktPublishing/`. Check them out!

Download the color images

We also provide a PDF file that has color images of the screenshots and diagrams used in this book. You can download it here: `https://packt.link/9bfGQ`.

Conventions used

There are a number of text conventions used throughout this book.

`Code in text`: Indicates code words in text, database table names, folder names, filenames, file extensions, pathnames, dummy URLs, user input, and Twitter handles. Here is an example: "Mount the downloaded `WebStorm-10*.dmg` disk image file as another disk in your system."

A block of code is set as follows:

```
html, body, #map {
 height: 100%;
 margin: 0;
 padding: 0
}
```

When we wish to draw your attention to a particular part of a code block, the relevant lines or items are set in bold:

```
[default]
exten => s,1,Dial(Zap/1|30)
exten => s,2,Voicemail(u100)
exten => s,102,Voicemail(b100)
exten => i,1,Voicemail(s0)
```

Any command-line input or output is written as follows:

```
$ mkdir css
$ cd css
```

Bold: Indicates a new term, an important word, or words that you see onscreen. For instance, words in menus or dialog boxes appear in **bold**. Here is an example: "Select **System info** from the **Administration** panel."

> **Tips or important notes**
> Appear like this.

Get in touch

Feedback from our readers is always welcome.

General feedback: If you have questions about any aspect of this book, email us at customercare@packtpub.com and mention the book title in the subject of your message.

Errata: Although we have taken every care to ensure the accuracy of our content, mistakes do happen. If you have found a mistake in this book, we would be grateful if you would report this to us. Please visit www.packtpub.com/support/errata and fill in the form.

Piracy: If you come across any illegal copies of our works in any form on the internet, we would be grateful if you would provide us with the location address or website name. Please contact us at `copyright@packt.com` with a link to the material.

If you are interested in becoming an author: If there is a topic that you have expertise in and you are interested in either writing or contributing to a book, please visit `authors.packtpub.com`.

Share Your Thoughts

Once you've read *Azure Containers Explained*, we'd love to hear your thoughts! Scan the QR code below to go straight to the Amazon review page for this book and share your feedback.

`https://packt.link/r/180323105X`

Your review is important to us and the tech community and will help us make sure we're delivering excellent quality content.

Download a free PDF copy of this book

Thanks for purchasing this book!

Do you like to read on the go but are unable to carry your print books everywhere? Is your eBook purchase not compatible with the device of your choice?

Don't worry, now with every Packt book you get a DRM-free PDF version of that book at no cost.

Read anywhere, any place, on any device. Search, copy, and paste code from your favorite technical books directly into your application.

The perks don't stop there, you can get exclusive access to discounts, newsletters, and great free content in your inbox daily

Follow these simple steps to get the benefits:

1. Scan the QR code or visit the link below

   `https://packt.link/free-ebook/9781803231051`

2. Submit your proof of purchase
3. That's it! We'll send your free PDF and other benefits to your email directly

Part 1: Understanding Azure Container Technologies

In this part, you will understand the different technologies Microsoft Azure provides for running and managing your containerized solution. You will also understand the requirements for the different technologies provided.

This part of the book comprises the following chapters:

- *Chapter 1, Azure Containers – Pleased to Meet You*
- *Chapter 2, Azure App Service – Running a Container Was Never That Difficult*
- *Chapter 3, Deploying Containers to Azure Functions*
- *Chapter 4, Azure Container Instances for Serverless Containers*
- *Chapter 5, Azure Container Apps for Serverless Kubernetes*
- *Chapter 6, Azure Kubernetes Service for Kubernetes in the Cloud*

1

Azure Containers – Pleased to Meet You

Whether you have been working with containers for years, are just getting started, or are just interested in what they have to offer in Azure, then this book is for you.

Microsoft Azure gives you a variety of options when it comes to using containers, each with its own traits, benefits, and challenges. When reading this chapter, you will get a refresher on containers and their benefits, what technologies are available to run them on Microsoft Azure, and finally, what use cases are the best fit for these different platforms. This chapter is a primer for the chapters to come and should provide you with enough background information to follow the in-depth discussion of these technologies later on.

In this chapter, we're going to cover the following main topics:

- Understanding containers and their benefits
- Getting to know containers in the Azure landscape
- Exploring technologies and use cases

Understanding containers and their benefits

Virtualization has been around for a long time, and we can go as far as to say that it is the duct tape that holds infrastructures together. Different platforms provided different features (think of **VMware**, **Hyper-V**, and **KVM**) but all had the same goal: **hardware virtualization**. We can now run multiple operating systems on a single piece of hardware, isolating them from each other and minimizing overhead. We got used to that. However, it did not answer all the questions or resolve the challenges we had. The world wanted to minimize overhead even more, add more flexibility, and have an answer to the comment, *it worked on my machine!*

Containers may have been around for much longer than traditional hardware virtualization in different forms, such as the **Unix chroot system** and **FreeBSD Jails**, but only became really popular in their current form with the introduction of **Docker** and the **Open Container Initiative** (**OCI**). The OCI was founded by Docker and other leaders in the container ecosystem in June 2015. It is an open source specification to ensure container images can work across multiple container runtimes.

Container technology these days is essentially what we would call **operating system virtualization**, where we package code, libraries, and the runtime into a **container image** and run it on top of an operating system, using a container engine such as Docker. To make a comparison with hardware virtualization, you can say that the container engine is the hypervisor for the containers. Of course, there is much more to it when you really get into the nitty and gritty of container technologies, but we don't need that level of understanding when navigating through the Azure container landscape. Let's see this in a visual representation.

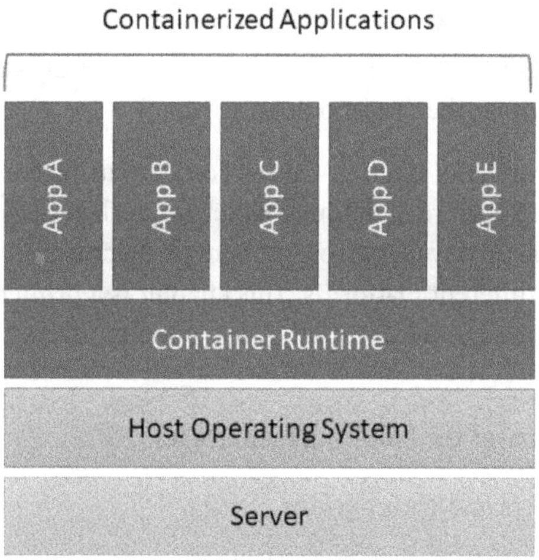
Containerized Applications
App A
App B
App C
App D
App E
Container Runtime
Host Operating System
Server

Figure 1.1 – An overview of containers

In the preceding diagram, you can see we still have a *server*, but the capacity is distributed more efficiently. Where traditionally we would run one application per (virtualized) server, with container technology, we can now run multiple isolated containers on a single operating system and minimize overhead even more.

> **Important note**
> When we talk about running a *container*, we are actually running an instance that is based off a *container image*. The container image actually contains all the code, libraries, and runtime but is more often referred to as a **docker container**. Throughout this book, we will use the term *container* when referring to a container instance that is created from a container image.

Container characteristics

These containers have specific characteristics and can be used in multiple ways, each use case coming with its own set of benefits. Let's take a look at these specific characteristics:

- Containers are lightweight.
- Containers are ephemeral.
- Containers contain everything required from an application perspective and all the -specific binaries that come from the underlying node **Operating System** (**OS**).
- Containers have strong default isolation.
- Containers contain the same content wherever you run them (working on everyone's computer).
- Containers can run on Linux or Windows.

That's a pretty interesting list, but those characteristics do come with some important side notes.

As containers are lightweight, they won't take up too many resources, and you can run hundreds of them on a single system. Instead of running hundreds of virtual machines, you are now running just a couple with hundreds of containers. At some point, we need to look at efficiently managing those.

As containers are ephemeral, this has consequences for your solution. We're talking stateless here. And, by default, containers have strong default isolation. This means, by default, two containers will not communicate with each other. That also has consequences for your solution and software architecture.

These consequences are not all that bad. In fact, if you play by the rules, you will end up with a more scalable, secure, and future-proof solution.

Container benefits

Maybe you could already tell from the previous paragraphs that there are definitely benefits to using container technologies:

- Containers contain everything you need to run your software.
- Containers are extremely scalable.
- Containers don't have much overhead.
- Containers are portable.
- Containers are faster than a traditional virtual machine.

That sounds very interesting (even for the financially minded people out there!). But what does it mean? Well, a container contains everything you need to run your software. Within your container image, you store the parts of the OS you need, the libraries you are using, and, of course, your code. That container image is stored in what we call a registry and can be used whenever you want to start your container. Whether that container is running in the cloud, on your local machine, or in your refrigerator (if it supports it), it will always have the same contents. It works on everyone's machine.

Having such a small footprint means that containers can be started really quickly but can also be scaled just like that. As containers also have significantly less overhead as compared to traditional configurations, instead of having to deploy multiple virtual machines to host multiple instances of your software, you can now do that by just running a number of small containers on the same machine.

> **Important note**
> A container registry is a repository that contains container images that can be *pulled* by other services to start an instance of a container. Microsoft Azure offers a service called **Azure Container Registry** that can be integrated into other Azure services.

It is very likely that you are not looking to run all these containers on traditional on-premises hardware, but you want to leverage the global scalability, cost efficiency, redundancy, and security that public clouds such as Microsoft Azure have to offer. And we're going to look into that right now!

Getting to know containers in the Azure landscape

You aren't reading this book because you want to run containers on **Amazon Web Service** (**AWS**), **Google Cloud Platform** (**GCP**), or on-premises. You're reading this because you are interested in the container landscape on Microsoft Azure. Let's take a look at that!

Microsoft Azure provides different technologies that support running container-based workloads. Each technology fits different use cases and has different behavior, and it's important that you select the right technology for the task at hand. There is no right or wrong solution; some are fit to run on enterprise-grade technologies such as **Azure Kubernetes Service** and some will do just fine on **Azure App Service** for containers. However, if you are a software company, it is unlikely that your solution is going to remain the same over several years. Business goals change, software architectures change, and public clouds change. Knowing what your options are, when to migrate, or when to reconsider a specific technology are key to successfully running containers in Microsoft Azure.

In this book, we will discuss, explain, and show multiple Azure container technologies and elaborate on their use cases. Let's briefly introduce these technologies:

- Azure App Service for containers
- Azure Functions for containers
- Azure Container Instances
- Azure Container Apps
- Azure Kubernetes Service
- Azure Container Registry

We will provide a brief overview of each technology in the next section and help you understand what they do and why.

Exploring technologies and use cases

Let's talk technology! Even though all these technologies can run one or multiple containers, they behave differently. In the next chapters of this book, we will deep-dive into each technology and its use cases. But first, let's introduce them!

Azure App Service for containers

Azure App Service were originally designed to host your web application or web APIs on a fully managed platform called Azure App Service. With the popularity of containers, the capability for running them on Azure App Service was introduced. Originally, only Linux containers were supported, but in 2020, Windows container support was added.

Getting up and running with containers on App Service requires you to point to a registry where your container image is located, and there you go!

At the time of writing this book, Web App for Containers officially only supports running single-container workloads. However, multi-container workloads are currently in preview.

If you are already using Azure App Service for other solutions and now need a single container workload to run on the same technologies that you are already used to, Azure App Service are worth exploring. In the next chapter, we will do a technical deep-dive into how you can get started, what other technical features Azure App Service for containers provides, and what you need to know before deploying them.

Azure Functions for containers

You might associate the term *serverless* with **Azure Functions**. And if your heart starts beating faster when you hear about these, you may well ask yourself why not run your containers on Azure Functions? To be fair, Azure Functions is not a platform designed to host your enterprise solution on containers. In fact, it's the other way around. Let's explain.

If you are familiar with Azure Functions, you might have noticed that from the outside, the management experience is very similar to Azure App Service. In fact, the technologies in the backend are very similar. The main difference is that Azure Functions is serverless while Azure App Service are not.

The main question is, why would you want to run your code in Azure Functions as a custom Docker container? The answer is quite simple and one of the benefits we have already discussed in a previous section: managing libraries, dependencies, and runtimes. Azure Functions only has certain runtimes available; with a custom container, you can use one that is not part of the default Azure Functions service. You could say that containers are an extension on top of Azure Functions and can be used when you are limited by the capabilities of Azure Functions itself. Where normally you would select a platform to run your containers on, you can now use containers to make the platform work better for you. Containers to the rescue!

Please keep in mind that, at the time of writing, running containers on Azure Functions is only supported for Linux and requires a premium or dedicated app service plan.

In *Chapter 3*, we will explore the technical capabilities of containers on Azure Functions and discuss how you would go about deploying these.

Azure Container Instances

Microsoft's first serverless container platform is **Azure Container Instances**. This platform is all about running containers and consuming resources on demand. Even though Azure Container Instances might look and sound like another average container platform, the key to success here is the available integrations with other Azure services.

Azure Container Instances is all about not managing the infrastructure. However, this also means that there is no control over the infrastructure. That is not a bad thing, but it is something that needs to be considered before deploying your containers to Azure Container Instances.

Let's get back to the integration part of things. As Azure Container Instances is *serverless* and *event-driven* by nature, we can trigger it from other Azure services. Perhaps you have a workflow defined in an Azure logic app and need to quickly spin up, run a container, and work with the outcome (a calculation for example); this can be configured in a matter of a few *clicks*. More complex tasks such as integration with Azure Functions, Azure Queue, and Azure Kubernetes Service are also supported.

And that is something we do need to mention – the integration with Azure Kubernetes Service. Let's say you have workloads that run on Azure Kubernetes Service but one of the characteristics of your solution is that there happen to be unpredictable bursts in resource requirements. This means we need more containers, more CPU, more memory, and we need it now! Azure Container Instances integrates with Azure Kubernetes Service to provide a form of *bursting*. If your Azure Kubernetes Service can't keep up with demand, you can have it automatically burst to Azure Container Instances for the duration of the *peak moment* and remove it again once it is no longer needed. All this and you are only billed per second once your Azure Container Instances instance is running.

We'd call that a perfect addition to an infrastructure that requires flexibility and resiliency.

In *Chapter 4*, we will dive into all that ACI has to offer.

Azure Container Apps

Where do we start? Well, at the time of writing, **Azure Container Apps** is still in preview and was announced at Microsoft Ignite 2021. It's essentially ACI on steroids or the new-found sibling of Azure Kubernetes Service. Azure Container Apps provides a series of Microsoft and community best practices wrapped into a single service that you can run containers on. Azure Container Apps is designed for organizations who need container orchestration but Azure Kubernetes might be something of an overkill.

Out-of-the-box Azure Container Apps comes with support for open source services such as **Kubernetes Event Driven Autoscaling** (KEDA), **Distributed Application Runtime** (**Dapr**), and a fully managed Ingress controller.

This means we can just focus on building the containers and run them, as long as we keep in mind to play by the Azure Container Apps rules. It's great to get accustomed to writing code fit for containers and following best practices without having to worry about infrastructure management. It's really a stepping stone to building enterprise architectures with containers on Azure.

Chapter 5, will be the main chapter where we will discover what Azure Container Apps has to offer.

Azure Kubernetes Service

First, we had the Azure Container Service, where we could choose between Docker Swarm, Kubernetes, and **Distributed Cloud Operating System** (**DC/OS**), but that service was retired. Kubernetes has been the de facto standard for container orchestration for some time, and Microsoft built a managed solution around that called **Azure Kubernetes Service**. The cool thing is that Microsoft follows the upstream Kubernetes project and adds additional services and integrations with Azure on top of that.

What you get are all the good things that Kubernetes has to offer but with a Microsoft Azure sauce on top of it. This means that everything you can run on Kubernetes, you can run on Azure Kubernetes Service.

Contrary to popular belief, it's not just for enterprises. Azure Kubernetes Service can already be leveraged for relatively small environments if done correctly.

Azure Kubernetes Service essentially makes running Kubernetes a lot easier. We no longer have to worry about managing and configuring etcd (a high-available key-value store for all cluster data), Kubernetes APIs, and the kubelet – that is now all done for us. Essentially, you get the control plane for free, but you are still responsible for upgrading your Kubernetes versions and your node images, including security patches. However, Microsoft Azure makes this process extremely easy by providing these features with the click of a button.

Azure Kubernetes Service is the answer to the limitations of the previously mentioned services. If your use cases go beyond what those services can do, the answer is usually Azure Kubernetes Service.

With the ability to scale to thousands of nodes, the extensibility of Kubernetes, and the solutions and add-ons that the cloud-native community provides, there is usually no question left unanswered and no challenge left unresolved. This might sound like a very big promise, but give us the time and opportunity to explain in *Chapter 6*.

Azure Container Registry

Those container images have to come from somewhere. The common technology across all the features mentioned in the previous paragraphs is **Azure Container Registry** (**ACR**). Technically, it doesn't host your containers, but it is the resource you will use to host or even build your container images.

You may even have heard of **Docker Hub**, which is a public container registry. ACR is basically the same but lives in Microsoft Azure. It can be both a public and private registry. It even has geo-replication support built in.

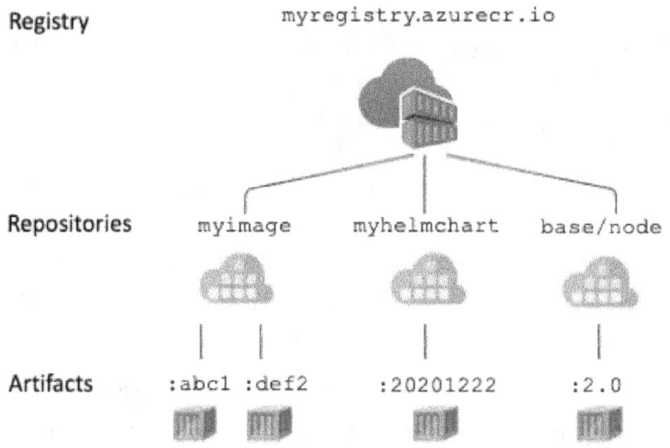
Registry
myregistry.azurecr.io
Repositories
myimage
myhelmchart
base/node
Artifacts
:abc1
:def2
:20201222
:2.0

Figure 1.2 – ACR elements

Let's break this diagram down and take a look at repositories, what they contain, and what additional features ACR provides in general.

Repositories

When we work with container images – for example, when we build a new one – the docker command will be something like `docker build imagename:tagvalue`. When you see *image name*, you can think of that as the repository name. Any container image you push to the container registry with the same image name but a different tag value will end up in the same repository. An example would be `docker build MyContainerApp:v1`.

You are also able to use namespaces. These are a helpful feature for you to easily identify related repositories. If we use the preceding example, `imagename:tagvalue`, we are able to add a namespace using a forward slash-delimited name. So, `imagename` could now look like `development/app1/imagename:tagvalue`. You can see that we have added `development/app1`. We can add this to another container image that falls under `app1` to help us identify that this container image is part of `app1`. One thing to note here is that even though we have the namespaces, ACR manages all container images independently. They are not a hierarchy.

> **Important note**
>
> When tagging container images, it is recommended to follow your versioning policy. Do not be dependent on the *latest* tag, as some services do not support that in continuous integration and continuous delivery/deployment scenarios.

ACR tasks

You're probably familiar with building container images using Docker on your local machine, but did you know ACR actually comes with a suite of features called **Azure Container Registry Tasks** (**ACR Tasks**) that allows you to build container images using the cloud? You are able to not only build Linux or Windows containers but also ARM containers too. ACR Tasks allows you to extend your development cycle to the cloud – for example, using ACR Tasks to build containers during a DevOps pipeline.

You are able to trigger ACR Tasks automatically in a few ways: through a source code update, a base image update, and a schedule. You are also able to trigger on demand, known as **quick tasks**.

Quick tasks

Most developers want to write code, build an application, and then test it locally before even doing a commit to source control. With containers, you would need a tool such as **Docker Desktop** installed to be able to build your container image locally. Docker Desktop is a great tool, but you are only able to build container images based off your hardware. So, if you are using a Windows machine, you are able to build a Windows image. If you install **Windows Subsystem for Linux** (**WSL2**), then you are also able to build Linux container images, but it also uses a lot of system resources. The more complex your solution becomes, the more powerful your local machine needs to be to build and run it. To overcome that, you can use ACR quick tasks to build the container image in the cloud. You are also able to run the container image inside ACR Tasks, but at the time of writing, it does not work well with complex container images, and you will have more success and flexibility testing your container on the target infrastructure.

If your DevOps build agents are not running on a machine or container that is capable of creating a container image, then offloading the building of the container image to the cloud using ACR quick tasks is an ideal solution. You just need to log in to your Azure subscription in your pipeline and use the `az acr build` command instead of `docker build`.

Currently in preview at the time of writing this book is the ability to build and push a container image directly from the source code, without a Docker file. This new feature uses an open source tool called **Cloud Native Buildpacks** (`https://buildpacks.io/`).

> **Important note**
>
> Note that DevOps build agents are not specifically Azure DevOps build agents, but in general, a large number of DevOps solutions (Jenkins, Octopus Deploy, and GitLab) support running on containers.

Automated tasks

You are able to connect your public or private Git repository and optionally branch in both GitHub and Azure DevOps to an ACR task. By default, when you configure an ACR task to monitor your Git repository, it will run with every commit. You are able to configure it to run on a Pull request as well. When you update code in your repository, the ACR task is triggered via a Webhook it creates and will build the container image and push it to the container registry ready for use. This is extremely useful when doing automated testing in your pipeline.

Container images, just like virtual machines, need to be kept up to date. Now you could do this manually, but that would mean you need to update your base images, then your main images, and so on, which is a lot of work. A base image is the starting image of your container image. It would normally be something like an Ubuntu version with perhaps some added applications. Then, your code is added on top to make your application container image.

ACR Tasks has your back. You are able to create a task that automatically detects when a base image has been updated in your registry or a public registry, such as Docker Hub. Once the task detects that the base image has been updated, it will then create a new version of your container image and push it to the correct repository.

You may need to run a maintainer task to clean up your repository of old container images or test a build and push it to your registry. For this, ACR Tasks has scheduled tasks. There's not much more we can say about them, apart from that they are really helpful when you need to remove old container images or feature build images, as the `purge` command comes with a filter option that uses a regex.

Multi-step tasks

You may have some requirements to test your application before it is pushed to a container registry. Multi-step tasks have you covered here. With multi-step tasks, build and push tasks are separated. You have the ability to create a task that can build your application container image and then run it. It can then build and run another container image that has your testing tools inside. This testing container will perform your tests against your running application. If they pass the tests, then the image can be pushed to the container registry in the next part. If they fail the test, the image is not pushed to the container registry.

Multi-step tasks allow you to have more granular control over image building and testing to ensure only good images are pushed to the container registry.

Summary

In this chapter, we have provided an overview of the services that Microsoft Azure has to offer when it comes to running containers. On a high level, you have learned what these services do and what their purpose is. We can run containers with a very small footprint on very accessible services such as Azure App Service and ACI, but we can also go for an enterprise-grade scenario by using Azure Container Apps or Azure Kubernetes Service.

What we have learned is that containers in general are not that different from traditional hardware virtualization; we're just doing things more efficiently.

There are many flavors, each with a different set of use cases. And even though you might already have a favorite service based on what we have described in this chapter, please be patient and let us guide you through each service while uncovering what they really have to offer, how they work, and what you need to do to get started.

In the next chapter, we will start by diving into the world of containers on Azure App Services and gradually move on to the more complex features in the chapters that follow. Now that you have been introduced to what container services the Azure landscape is comprised of, let's get technical.

2
Azure App Service – Running a Container Was Never That Difficult

If you are new to Microsoft Azure or have been using it for some time, you have probably come across **Azure App Service**. Azure App Service is a **Platform-as-a-Service** (**PaaS**) offering designed to run web applications, including websites and web APIs.

Back in 2017, Microsoft released *Azure App Service on Linux* and *Web App for Containers*. App Service was already quite popular in the Windows world, but it now also supports Linux and containers! As the name suggests, at launch, it only supported Linux container images. In 2017, there were few use cases for Windows containers as the operating system still required some tweaks to be ready for containerization. Over the years, the popularity of Windows containers increased; now, you can run Windows container images too. Containers matured and so did the platforms supporting them.

If you investigate the available technologies with container support on Microsoft Azure, your first thought might be that App Service might not be the first go-to platform to run your container images on. It may even feel like the feature was *shoe-horned* in or not *native* to App Service. But hear us out – there are advantages to running containers on Azure App Service. It's not about picking the most popular technology; it is about picking the right technology for you.

In this chapter, we will cover the following topics:

- Understanding Azure App Service for Containers
- Deploying containers to Azure App Service
- The pros and cons of deploying containers on Azure App Service

Understanding Azure App Service for Containers

Before we explore Azure App Service for Containers, we need to understand Azure App Service. This offering helps you utilize the power of Microsoft Azure in your application. Out of the box, you not only have better security, load balancing, autoscaling, and even automated management – you also have built-in integration with Azure DevOps and GitHub for continuous deployments. Azure App Service is also described as a *fully managed platform*, where Microsoft will take care of infrastructure and management. You will focus on building functionality (your solution).

Azure App Service for Containers runs container images rather than code natively – you even get all the same benefits listed in the previous paragraph.

In the backend, Azure App Service runs the Docker runtime to host your container images, but this is completely abstracted away from you, so you do not have to worry about keeping it up to date with security patches and new major versions. It even comes with some built-in images that you are free to use, though you can use any custom image from any container registry. You can find a list of the built-in images by using the **Azure Command-Line Interface** (**Azure CLI**) and running the following command:

```
az webapp list-runtimes --linux
```

In the next section, we will guide you through the steps that are required to create an Azure App Service and run your container on it. For this, we will use **Azure Cloud Shell** via the Azure portal. By doing it this way, you will familiarize yourself with the Azure portal. Moreover, Azure Cloud Shell ensures that all the required tooling to follow these steps is present and that you are logged in with the correct permissions.

> **Important note**
>
> It is also possible to use the Azure CLI on your local machine. For this, you need to install the Azure CLI, which is available for different operating systems at `https://docs.microsoft.com/en-us/cli/azure/install-azure-cli`.

Everything is better with an example. Before we start deploying, let's look at an actual use case and see whether we can perform the tasks to meet the requirements.

> Your company provides an eCommerce solution to the market. Your solution is built on several different Microsoft Azure technologies.
>
> Your product manager informs you that they have purchased some licensed code to help them process billing statements. This code is an extension of the existing e-commerce platform. You are already using Azure App Service but the code provided does not use a supported runtime for Azure App Service. You already have automation built around monitoring, scaling, and deploying App Service, and not using them would be considered deviating from your standard. It seems like the best way to stick to your standards and get the solution up and running is to containerize the billing statements solution and run it on Azure App Service until it is modernized to a more suitable runtime that is supported.
>
> It is your job to get this up and running.

Let's see if we can deploy App Service and the billing statements API.

Deploying containers to Azure App Service

In the previous section, you learned what Azure App Service for Containers is. Now, it is time to create your first container. To achieve this, we need a few things:

- Azure Cloud Shell with the Azure CLI to run our commands
- A resource group as a place for our resources to live
- An App Service plan for hosting our App Service
- A web app to host our container image
- The container image

This sounds like a lot, but luckily, a lot of these resources can be created just by running a couple of commands. It is, however, important to understand the *layers* we just described in the preceding list.

First, we need a resource group since every resource in Azure needs a place to live. The resource group is the logical container in which we can group our resources, configure **role-based access control** (**RBAC**), or use a scope for our Azure policies.

Then, we need a form of *hosting*. For that, we will deploy an App Service plan where we define and allocate what capacity we need (SKU). Within this App Service plan, we can host one or multiple web apps. In this example, we will be deploying a single web app, but you are free to repeat the steps provided and deploy multiple ones. This web app will contain the runtime (Docker) for starting our container and running our solution.

Let's start with Azure Cloud Shell. We are going to use Azure Cloud Shell via the Azure portal. Go to the Azure portal at `https://portal.azure.com`. Once you've logged in, click on the **Cloud Shell** icon near the search bar, as shown in the following screenshot:

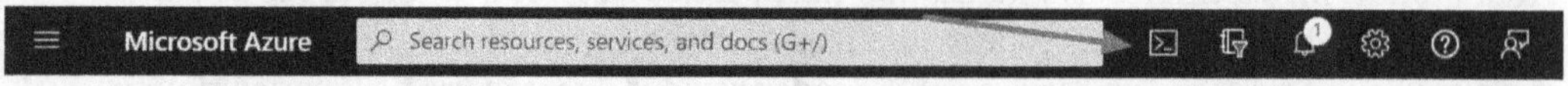
Microsoft Azure
Search resources, services, and docs (G+/)

Figure 2.1 – Clicking the Cloud Shell icon to open Azure Cloud Shell

If this is the first time you have used Cloud Shell, you will be asked to choose between **Bash** or **PowerShell**. Go ahead and choose **PowerShell** as we will be using PowerShell throughout this book:

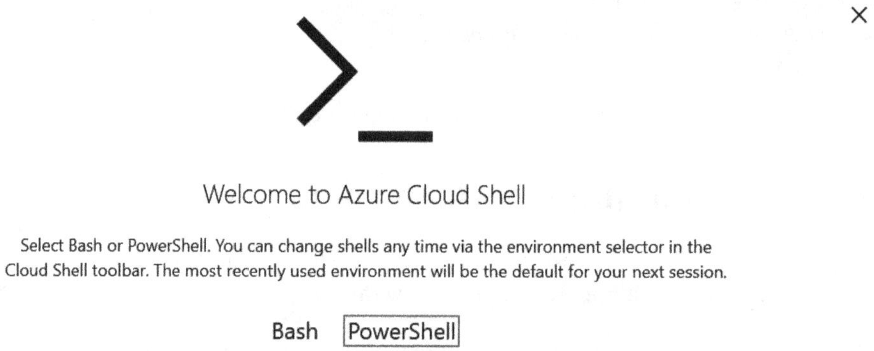
Welcome to Azure Cloud Shell
Select Bash or PowerShell. You can change shells any time via the environment selector in the Cloud Shell toolbar. The most recently used environment will be the default for your next session.
Bash
PowerShell

Figure 2.2 – Selecting the PowerShell option

Now, you will be asked to create a storage account. Confirm the correct subscription and click on **Create storage** to continue:

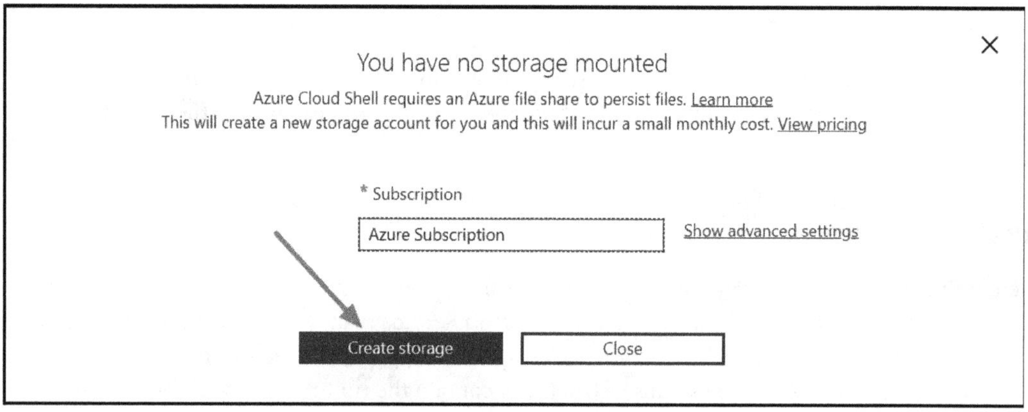
You have no storage mounted
Azure Cloud Shell requires an Azure file share to persist files. Learn more
This will create a new storage account for you and this will incur a small monthly cost. View pricing
* Subscription
Azure Subscription
Show advanced settings
Create storage
Close

Figure 2.3 – Creating a new storage account for Cloud Shell

This will automatically create a new storage account attached to Cloud Shell. You can use this to store scripts and code that you may need later.

> **Important note**
>
> If you receive a warning message saying it failed to mount the Azure file share, it is because the storage account has not fully completed provisioning yet. Just restart your Cloud Shell; it should start as expected.

We are going to use the Azure CLI to create the resources in Azure. As mentioned previously, a huge benefit of using Azure Cloud Shell is that the Azure CLI comes preinstalled and is regularly updated, making sure that you can use the latest and greatest features.

Before you can create any resource in Azure, you need a resource group. To create a resource group, you need to type the following command:

```
az group create `
--name "rg-appserv-containers"  `
--location "west europe"
```

> **Important note**
>
> Notice the use of backticks (`) to separate longer commands over multiple lines to help with readability. If you are reading a digital version of this book, you can copy and paste the commands easier. If you are reading a physical copy of this book, you can just type the commands on one line by ignoring the backticks. If you are using **Bash** instead of **PowerShell**, you will need to replace the backticks with a backslash (\).

To verify that the resource group has been created, type the following:

```
az group show --name "rg-appserv-containers"
```

You should see something similar to the following:

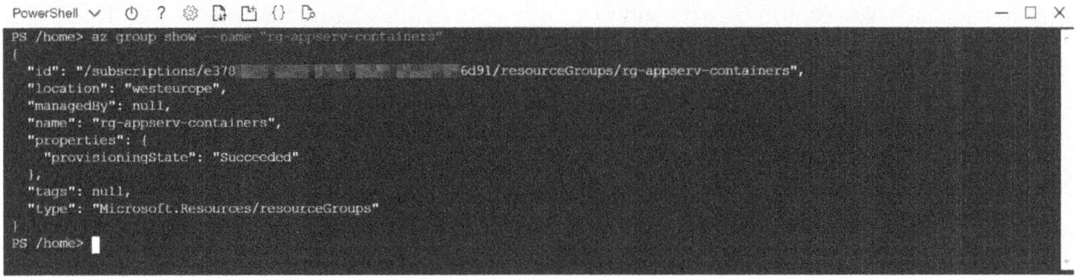
PowerShell
PS /home> az group show --name "rg-appserv-containers"
{
"id": "/subscriptions/e378 6d91/resourceGroups/rg-appserv-containers",
"location": "westeurope",
"managedBy": null,
"name": "rg-appserv-containers",
"properties": {
"provisioningState": "Succeeded"
},
"tags": null,
"type": "Microsoft.Resources/resourceGroups"
}
PS /home>

Figure 2.4 – Output of the az group show command

Now that we have a resource group, we can create the Azure App Service plan. To do this, use the following command:

```
az appservice plan create `
--name "containersplan" `
--resource-group "rg-appserv-containers" `
--location "west Europe" `
--is-linux `
--sku s1 `
--number-of-workers 1
```

> **Important note**
>
> If you want to run a Windows Containers App Service, you must replace `--is-linux` with `--hyper-v`.

Let's break this command down. `az appservice plan create` is telling Azure that we will be creating that little hosting part called `App Service Plan`. We provided it a name with the `–name` switch, and we told it where to live with `--resource-group`. We also provided a location, `west europe`, and told the App Service plan that its operating system is Linux with the `--is-linux` switch.

As mentioned previously, we need to allocate capacity – that's what we did with `–sku`. Here, we selected `Standard 1 SKU`. Finally, we needed to decide on the instance count (how many of those S1 instances we need). In this example, we only need one.

> **Important note**
>
> The maximum number of workers (or instance count) differs per SKU. For instance, a premium tier App Service plan can have more instances than a standard tier App Service plan.

After a few seconds, you should have an output similar to the following:

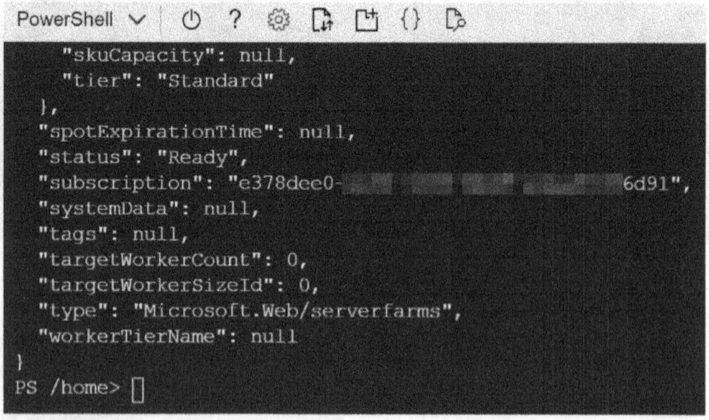
PowerShell
"skuCapacity": null,
"tier": "Standard"
},
"spotExpirationTime": null,
"status": "Ready",
"subscription": "e378dee0-[redacted]6d91",
"systemData": null,
"tags": null,
"targetWorkerCount": 0,
"targetWorkerSizeId": 0,
"type": "Microsoft.Web/serverfarms",
"workerTierName": null
}
PS /home>

Figure 2.5 – Output of the Azure App Service plan create command

With the App Service plan created, we can deploy a container to the App Service plan. For this, we will be creating an Azure web app inside the App Service plan. To do this, use the following command, though be aware that `--name` needs to be globally unique:

```
az webapp create `
--name "containersinazure" `
--resource-group "rg-appserv-containers" `
--plan "containersplan" `
--deployment-container-image-name "whaakman/container-
demos:billingstatementsv3"
```

Here, we told Azure that we want to create that web app within the App Service plan that we just created by defining the resource group and the name of the plan that the app will be running in.

In your Cloud Shell, you should have an output similar to the following:

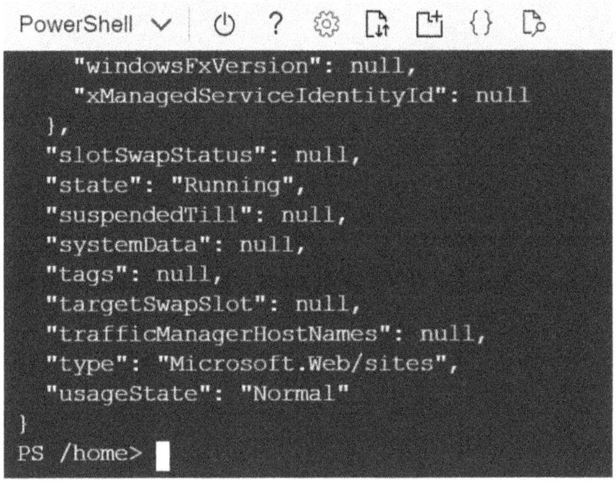
PowerShell
"windowsFxVersion": null,
"xManagedServiceIdentityId": null
},
"slotSwapStatus": null,
"state": "Running",
"suspendedTill": null,
"systemData": null,
"tags": null,
"targetSwapSlot": null,
"trafficManagerHostNames": null,
"type": "Microsoft.Web/sites",
"usageState": "Normal"
}
PS /home>

Figure 2.6 – Output of the az webapp create command

Let's go over the last part of the command to explain how we have deployed a container. The first three switches, `--name`, `--resource-group`, and `--plan`, are the same as you would use when deploying a regular web app to an App Service plan and are mandatory switches. The `--deployment-container-image-name` switch is the key here. With this switch, we are telling Azure that we will be running a container and we specify a container image to deploy to the web app. In this case, we have used the `nginx` container image from Docker Hub. You may have noticed the command only says *nginx*. This is because, with most container systems, Docker Hub is always implied. You could put the full path to your container image. For example, if it was stored in Azure Container Registry, you could supply the following switches to pass in authentication:

```
--docker-registry-server-user
--docker-registry-server-password
```

Seeing it in action

If you followed along with the code in the previous section, you should now have an *nginx* container running in a newly created web app. But how do we access it to see whether it is working? For this, we are going to use the Azure portal.

In the Azure portal, use the search bar at the top and search for *web app*. Under the **Services** section, click on **App Service**:

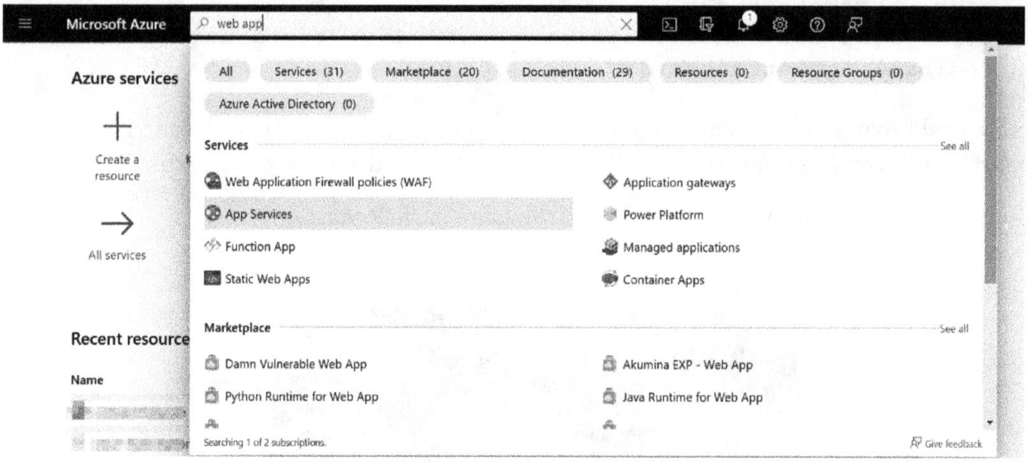
Microsoft Azure
web app
Azure services
Create a resource
All services
Recent resource
Name
All
Services (31)
Marketplace (20)
Documentation (29)
Resources (0)
Resource Groups (0)
Azure Active Directory (0)
Services
See all
Web Application Firewall policies (WAF)
Application gateways
App Services
Power Platform
Function App
Managed applications
Static Web Apps
Container Apps
Marketplace
See all
Damn Vulnerable Web App
Akumina EXP - Web App
Python Runtime for Web App
Java Runtime for Web App
Searching 1 of 2 subscriptions.
Give feedback

Figure 2.7 – Searching for web apps within the search bar

Here, you will see your newly created web app. Click on it to go to the overview blade:

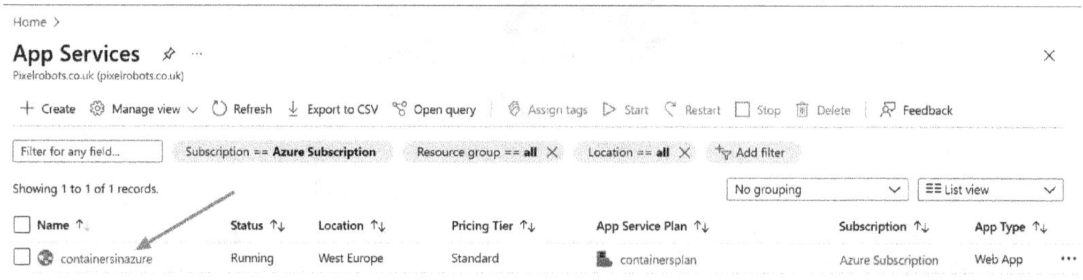
Home >
App Services
Pixelrobots.co.uk (pixelrobots.co.uk)
Create
Manage view
Refresh
Export to CSV
Open query
Assign tags
Start
Restart
Stop
Delete
Feedback
Filter for any field...
Subscription == Azure Subscription
Resource group == all
Location == all
Add filter
Showing 1 to 1 of 1 records.
No grouping
List view
Name
Status
Location
Pricing Tier
App Service Plan
Subscription
App Type
containersinazure
Running
West Europe
Standard
containersplan
Azure Subscription
Web App

Figure 2.8 – Clicking the web app to get to the overview blade

The overview blade is where you can get a quick view of your deployed container. It displays the resource group and location it is deployed in. It also shows you its current status. In the following screenshot, you can see that the container's current **Status** is **Running**:

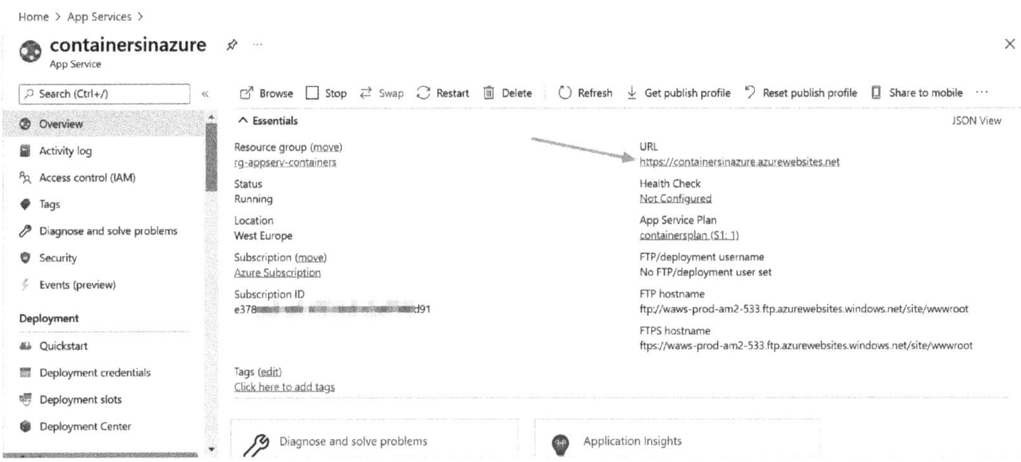
Home > App Services >
containersinazure
App Service
Search (Ctrl+/)
Browse
Stop
Swap
Restart
Delete
Refresh
Get publish profile
Reset publish profile
Share to mobile
Overview
Activity log
Access control (IAM)
Tags
Diagnose and solve problems
Security
Events (preview)
Deployment
Quickstart
Deployment credentials
Deployment slots
Deployment Center
Essentials
JSON View
Resource group (move)
rg-appserv-containers
Status
Running
Location
West Europe
Subscription (move)
Azure Subscription
Subscription ID
e378[redacted]d91
Tags (edit)
Click here to add tags
URL
https://containersinazure.azurewebsites.net
Health Check
Not Configured
App Service Plan
containersplan (S1: 1)
FTP/deployment username
No FTP/deployment user set
FTP hostname
ftp://waws-prod-am2-533.ftp.azurewebsites.windows.net/site/wwwroot
FTPS hostname
ftps://waws-prod-am2-533.ftp.azurewebsites.windows.net/site/wwwroot
Diagnose and solve problems
Application Insights

Figure 2.9 – Overview blade showing the deployed container URL

You will also notice that there is a URL. This is the live URL of your deployed container. Now, we can invoke a call to our URL that contains an endpoint called `/BillingStatement`. By calling that endpoint, we should receive multiple billing statements, indicating that everything is up and running.

Now, let's execute the following command:

```
Invoke-RestMethod -Uri "https://containersinazure.
azurewebsites.net/BillingStatement"
```

As App Services are continuously running the container should be started and the command should return a result almost instantly.

As you can see, the container returns the billing statements, indicating that this was a success. Now, we can go back to our product manager and share the good news.

> **Important note**
>
> If you want to try running multiple containers in a single web app for containers (a feature that is in preview at the time of writing), please refer to the documentation at `https://docs.microsoft.com/en-us/azure/app-service/tutorial-multi-container-app`. The requirements are similar to what we described earlier and merely require adding additional parameters to Azure CLI commands for deploying the container image.

In this section, you learned how to create an Azure App Service plan, create an Azure web app for containers, and deploy the public nginx container hosted on Docker Hub. You did this by using Azure Cloud Shell and the Azure CLI. The examples we covered here are not fit for production and should be used as a starting point to learn how to deploy containers to Azure App Service.

We don't want to end up with an invoice for unused resources at the end of the month, so let's delete the resources that we created. We can do this by deleting the resource group using the following command:

```
az group delete --name rg-appserv-containers
```

With that, you have built an Azure App Service plan and deployed an nginx container image from Docker Hub to it. Now, let's go over some of the pros and cons of doing so.

The pros and cons of deploying containers on Azure App Service

As you learned in this chapter, it does not take much to deploy a container to Azure web apps. However, as it goes with technology, everything comes with advantages and disadvantages. Where the ease of use might be a big advantage, it doesn't always have to be. Let's discuss these advantages, limitations, and how they can impact your solution both short and long term.

Pros

Microsoft did not build support for containers on App Service just so we could write about it. They are there for a reason and they bring something to the table. Let's discuss the advantages!

Simplified scaling

Using Azure App Service means we have access to the Azure App Service plan scaling mechanisms. In short, we have two options:

1. Scaling vertically by upgrading to a different SKU (S1 -> S2 or P1) and essentially allocating more capacity for our solution
2. Scaling horizontally by adding multiple instances (workers), which means we end up with multiple instances/copies of our solution

Whereas any solution would support vertical scaling, this is different when we look at horizontal scaling. You must keep in mind that horizontal scaling requires your solution to support instances being added and removed on the fly. One moment, your app could be running on four instances, while another moment it could be running on two. Things such as state, session management, and startup times are trivial to the success of horizontal scaling.

However, if your solution supports horizontal scaling, you can use the built-in features of Azure App Service to automate this. You can do this based on a metric (for instance, CPU or memory usage) or based on a schedule.

These features come out of the box and take almost no time to implement. There is a small downside to using this mechanism as it is only supported on Azure App Service. Moving your containerized solution to a different technology means you will need to implement different scaling mechanisms.

App Service is a fully managed platform

As mentioned previously, App Service is a fully managed platform. This means that Microsoft takes care of a lot of the traditional infrastructure management that you would normally have to do yourself in a virtual machine scenario. With Azure App Service, you no longer have to deal with updating operating systems, installing security patches, or installing and configuring complicated devices/solutions for load balancing, high availability, and disaster recovery.

However, you are running a container image within that fully managed platform. As mentioned previously, you are building the required functionality. This means that in this case, you are building/developing whatever is inside the container. Microsoft cannot manage what is inside the container. What we're speaking of here is what we call a *shared responsibility*. Microsoft performs a lot of the management at an infrastructure level, and you manage what is inside the container.

If you build your own container image, that means you are responsible for updating the container image (installing the security patches and adding new dependencies).

So, what is the advantage here? You only have to worry about the container and what's inside it. The rest is managed for you.

Any runtime

Let's say you need a runtime that is not available on Azure App Service (or any other Azure technology that you have access to). Say you want to run a legacy runtime, or you are feeling adventurous and want to try out .NET 7 Preview. This is when containers are your go-to technology. You may already be running large parts of your solution in Azure App Service, but you want to test and discover how that solution would run on .NET 7 Preview and rule out any unsupported code that you might have.

Simply containerizing the current solution (which can usually be done out of the box using Visual Studio) and deploying it to an App Service plan, as described earlier in this chapter, will help you achieve that.

That doesn't mean you will always have to run that solution in a container. You might use the container to test the new runtime and know that when .NET 7 hits general availability, you can safely deploy and run your solution using the new runtime.

Azure DevOps and GitHub integration

Azure App Service provides tight integration with DevOps platforms such as Azure DevOps and GitHub. Since we are running a container on Azure App Service, we now have access to those features.

These features include the following:

- Post-deployment Webhooks
- GitHub Actions
- Azure Pipelines
- Integration with Azure Container Registry tasks

You can leverage these features via the **Deployment Center** feature of Azure App Service:

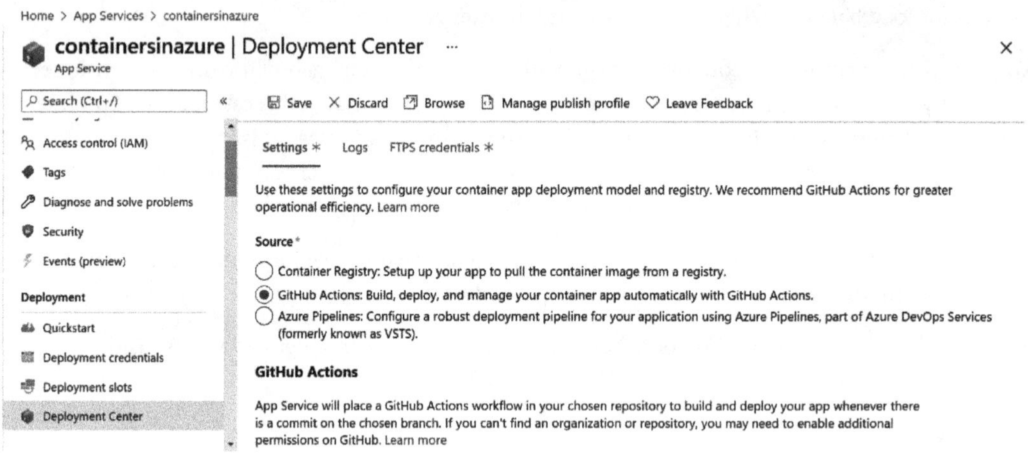
Home > App Services > containersinazure
containersinazure | Deployment Center
App Service
Search (Ctrl+/)
Save
Discard
Browse
Manage publish profile
Leave Feedback
Access control (IAM)
Tags
Diagnose and solve problems
Security
Events (preview)
Deployment
Quickstart
Deployment credentials
Deployment slots
Deployment Center
Settings *
Logs
FTPS credentials *
Use these settings to configure your container app deployment model and registry. We recommend GitHub Actions for greater operational efficiency. Learn more
Source *
Container Registry: Setup up your app to pull the container image from a registry.
GitHub Actions: Build, deploy, and manage your container app automatically with GitHub Actions.
Azure Pipelines: Configure a robust deployment pipeline for your application using Azure Pipelines, part of Azure DevOps Services (formerly known as VSTS).
GitHub Actions
App Service will place a GitHub Actions workflow in your chosen repository to build and deploy your app whenever there is a commit on the chosen branch. If you can't find an organization or repository, you may need to enable additional permissions on GitHub. Learn more

Figure 2.10 – The Deployment Center blade showing integration with GitHub

The **Deployment Center** feature allows us to link a source (for example, GitHub), where we can select where our code/container configuration is hosted, and have Azure generate the GitHub workflow for GitHub Actions automatically. By simply clicking the buttons in **Deployment Center** and authorizing access to GitHub, we can set up our **Continuous Integration/Continuous Delivery** (**CI/CD**) workflow within minutes.

This will create a workflow file in your GitHub repository. You are free to fine-tune it.

Windows and Linux

If you have been using containers, you have likely been using Linux containers. In the past few years, Windows containers have grown in popularity and are now also supported on Containers for Azure Web Apps. This has brought a whole new set of use cases to the table.

Looking at what happens when organizations take the path of *app modernization*, a lot of code probably requires refactoring, rearchitecting, and testing. If your solution is split up into multiple services (or you're looking to go that route), you will likely encounter a scenario where most of your solution can run in a Linux container but there's that one service that cannot. Reasons for this could be technical, but it could also be that you simply do not have the development resources (time, finances, or people) to achieve that short term. Those legacy configurations usually do work in a Windows container. This means you will end up with a landscape that contains a combination of both Windows and Linux containers.

> **Important note**
> In general, it is preferred to run containers using Linux as the Linux operating system requires less overhead, contains a small footprint (smaller image sizes), and operates faster than the average Windows container.

An important addition that we need to be aware of is that operating systems are selected upon deploying the App Service plan. In the previous examples, we have seen that distinction being made by using a single parameter when running the `az appservice plan create` command:

- For Linux: `--is-linux`
- For Windows: `--hyper-v`

This means that if we want to run both Windows and Linux containers (or just Windows and Linux app services), we need two different App Service plans. Keep in mind that the App Service plan is what you are paying for.

Cons

Where there are advantages, there are also disadvantages. Let's take a look!

Scaling methods

We mentioned scaling as an advantage, but we also discussed what the overall limitation is. Even though the horizontal and vertical scaling features in Azure App Service are amazing, they are not the mechanisms you would use in that sort of way in a more cloud-native world.

If we look at containerized solutions, or even better, a solution that consists of multiple containers, we normally do not use any features such as *vertical scaling*. Containers by nature are designed to support horizontal scaling, not just for scaling but also for running multiple instances of a container. Your solution supporting this means you have lots of flexibility when it comes to updating and upgrading both your solution as well as the underlying infrastructure. We want to follow the concept of *infrastructure as cattle*. Containers come, containers go, and we don't want to focus on what happens to a single instance of a container – we want orchestration to take care of that.

That's not to say that this is not possible on Azure App Service – we just discussed that horizontal scaling is supported. However, the Azure App Service plan itself just provisions an additional *instance*, unaware of the runtime. Azure App Service does not take into account that it is a container running on App Service. On the other hand, if you wish to run and manage more complicated scenarios where you run multiple containers, you want your underlying platform to be aware of what's running. This is what we see in the platforms more focused on containers such as Azure Container Apps, Azure Container Instances, and Azure Kubernetes Service.

That doesn't mean you should not use Azure App Service, but it is an important limitation to understand if you are planning to grow your container estate on the Azure platform.

Limited "do-it-yourself" capabilities

Azure App Service was not originally built for running containers, which we can see in the support for what we call *do-it-yourself* features. Whereas the regular Azure App Service runs the code directly and has tight integrations with solutions such as Application Insights and the Kudu console itself, Web Apps for Containers does not. The infrastructure details and logging are still accessible but if we want to debug our container, we need to modify our solution to provide the right type of logging to the Azure platform.

Other services provide integrations with features such as Container Insights, which is built to help you understand and monitor the behavior of your containers. However, as we mentioned previously, Azure App Service is not natively built to support containers, so we are limited in our do-it-yourself toolbox.

The question is, is this a disadvantage? You could argue that if you have already decided to run your container on Azure App Service, you probably have a small solution or want to test something, as discussed earlier in this chapter. If that is the case, do you need to get into the nitty and gritty of the backend technologies? If you want that, you will likely choose a different technology in Azure to host your containers on.

Currently single container

At the time of writing, Web Apps for Containers only supports running a single container. If your solutions consist of multiple containers (for example, a web app and an API), then these would require two different containers running on two different web apps.

Mind you, these can still run in the same App Service plan. This would not incur additional costs but it would take up additional capacity of the App Service plan.

Docker Compose (in a multi-container scenario)

For multi-container scenarios, at the time of writing, the use of Docker Compose is in preview. The scenario we just described, where you could run a web app and an API, is possible using this technology.

The downside, however, is that you are doing this with Docker Compose. It is fast, easy, and can be generated using Visual Studio, for example, but it is not something that's used that often anymore in *bigger scenarios*. Where Kubernetes became the de facto standard for managing containers, the majority of the community and businesses are now using Kubernetes Pods, which contain one or multiple containers. We will dive deeper into those later in this book, but it's important to understand that while a Pod may look similar to what we do with Docker Compose, it is very different. The advantage of this disadvantage? It's not rocket science to translate a Docker Compose file into a Pod definition for use with Kubernetes. Spent time is not lost!

Only suitable for web solutions (websites and APIs)

When we start running more complicated scenarios, we don't just want to run a website and a set of backend APIs. Maybe we want to introduce some caching via Redis Cache or maybe even an instance of Rabbit MQ. For Redis Cache, you could opt to use the default Azure implementation, known as Azure Cache for Redis. For Rabbit MQ, we don't have those options, so we would run it in a container.

This is where we have a challenge. Rabbit MQ runs on port `5672` (`5671` for SSL). Azure App Service only exposes ports `80` and `443`. If you wanted to run Rabbit MQ within an Azure App Service, then the only option would be to use Docker Compose and allow your other container to access the RabbitMQ container over the internal network. Any access from the outside is not supported for the reason that we just discussed: App Service only exposes ports 80 and 443.

For many scenarios, this would not be an issue, but it is a limitation you cannot overcome by changing the configuration.

As you can see, there is a use case for running containers in Azure App Service if you want to test that new unavailable runtime or need to run a legacy runtime on App Service; this is one of the best use cases. You also don't have to worry about platform management as the promise of App Service is a *fully managed platform*. All those awesome features such as the Deployment Center, horizontal and vertical scaling, and integrations with other Azure features are there out of the box.

The question here is, how much would the disadvantages impact you? And if you stick to the scenario we just described and you don't want to bother with managing the platform or you are already using Azure App Service and are used to it, then yes, those disadvantages may not even impact you.

The disadvantages start becoming noticeable when you start outgrowing Azure App Service. When your solution becomes bigger and more complex and your business has more demanding requirements when it comes to scaling, resiliency, and flexibility, that is when you start experiencing them.

It's important to understand that no time or resources are lost when this happens. You have already containerized your solution and as your business is growing, you probably have more financial momentum to move to a technology that is feature-rich but does come with higher costs. It is all about managing your business case. After all, you wouldn't buy a seven-seat car if you are not planning to have children yet.

Summary

In this chapter, you learned what Azure App Service and Azure App Service for Containers are and then created the Azure resources you need using Azure Cloud Shell. You started by creating a resource group and then a Linux-based Azure App Service plan. Inside this Azure App Service plan, you created a web app and set the deployed container image to nginx from Docker Hub.

After the deployment, you used the Azure portal to search for the newly created web app and used the overview blade to find the URL of your running container. Then, you navigated to view the nginx welcome page.

Next, you learned about the pros and cons of using Azure App Service to deploy your container workloads. We discussed a couple of advantages and disadvantages, some with a smaller and some with a bigger impact. You now know what Azure App Service for Containers is and how to create a simple Azure App Service plan and web app that runs a container image from Docker Hub. As mentioned in this chapter, the examples in this chapter are not production-ready but are designed to give you a taste of what is possible.

In the next chapter, you will learn about Azure Functions, how to run containers using it, and the pros and cons of using it to host your container workloads.

3
Deploying Containers to Azure Functions

Who doesn't love **Azure Functions**, right? They are extremely flexible and can be used in many different ways. With the increasing popularity of serverless runtimes and the benefits they bring, Microsoft decided to launch its own serverless proposition called Azure Functions in 2013. Like Azure App Service, Azure Functions is officially a part of the **Platform as a Service** (**PaaS**) portfolio and even similar in terms of looks. But, if we really look into their capabilities, they are quite different! Where both run on App Service plans, support multiple languages, and can run containers, that is where most similarities end. If we look at the inner workings, Azure Functions is considered *serverless* and Azure App Service is not.

Serverless is all about *on-demand compute*. So yes, there are still servers (compute), but they are not running by default. In fact, most components are switched off until they are required to execute. We're now focusing on just writing the code that matters, the logic we need.

Serverless is a big promise and in practice, we see that it is being used more often. Azure Functions are turned off by default, only run when we ask them to, and we only pay for them when they are actually consuming compute.

That really sounds like something different from a container. So, how would containerized applications fit in here? To be fair, Azure Functions is not a technology you would pick to run the majority of your containers on. While all other technologies in this book have at least one or two use cases where they are the recommended technology for your containerized solution, with Azure Functions, it works exactly the other way around.

If we are unable to do something in our Azure Functions landscape, we can extend it without introducing completely different technologies and just run a container alongside our other Azure Functions. In the following paragraphs, we will focus on how this would work in practice and how you can achieve this yourself.

In this chapter, we're going to cover the following main topics:

- Understanding Azure Functions for containers
- Deploying containers to Azure Functions
- Pros and cons of deploying containers on Azure Functions

Understanding Azure Functions for containers

As we discussed in the introductory chapter of this book, containers are like packages containing everything it needs to run the solution. All we need is a container runtime. Azure Functions can provide that!

Normally, when you look at the capabilities of Azure Functions, you might quickly determine that the sky is the limit. But, let's picture the following scenario. You build your solution using Azure Functions, and you have about 20–30 Azure functions. Pretty awesome! But, you have that one runtime or that one specific task that just cannot run on Azure Functions. That is when containers come in.

Before we start doing some container magic, let's first take a look at Azure Functions in general. As we mentioned in the introduction, Azure Functions is serverless and only consumes compute when it is executed. So, how do you run it?

Azure Functions comes with *triggers* and *bindings*. Basically, triggers are what make an Azure function execute, and bindings are what the Azure function integrates with (a queue, for example). You can just run Azure Functions based on a time trigger every couple of minutes or hours. Or, you can configure a trigger that requires a payload (an HTTP trigger, for instance). When using the HTTP trigger, you perform an HTTP request to the function endpoint and the body or parameters you provided are sent to the function for processing. The Azure function executes, performs whatever logic you built, and then shuts down again.

The serverless part in Azure Functions is in the App Service plan (remember those from the previous chapter?). For Azure Functions, we can choose the *Consumption plan*. This makes Azure Functions truly serverless. Within the Consumption plan, the compute is only allocated when we trigger the function and is deallocated briefly after the Azure function has been completed.

Deallocation of the compute also means that all state (local storage and memory) is lost. So, we need to remember this: never store state in an Azure function. Additionally, the maximum run duration of an Azure function is 10 minutes. After those 10 minutes, the Azure function will stop processing and deallocate.

That comes with a downside. That downside is called the *cold start*. Because resources are not allocated, everything needs to be spun up when the Azure function is triggered. All code and dependencies are loaded when the Azure function is executed after not being used for a couple of minutes (as it then returns to a state of *cold*). That means the more dependencies you have, the longer it takes for the Azure function to start.

If you are already using Azure Functions in the Consumption plan and you run into any of these limitations, or the cold start is a problem for one or two of your functions, there is a solution! Microsoft Azure provides you with alternatives when it comes to the plans you are choosing from, as follows:

- Consumption plan
- Premium plan
- Dedicated plan

Each plan comes with a number of characteristics that are worth mentioning.

- **Consumption plan**: This is the plan we just discussed in the previous paragraph. It comes with automated scaling, and we only pay for the resources when they are running. When multiple calls are made, the plan can scale the number of Azure Functions instances automatically. This is great for dynamic scenarios with a high load. But, if the Azure function is deallocated, we do have that cold start.
- **Premium plan**: The Premium plan is a bit different. You still have the same scaling mechanisms, but the function is hosted on *pre-warmed workers*. That means resources are permanently allocated. You could say, "That is very similar to Azure App Service," and you are (almost) right. Having those resources allocated permanently means we don't have to deal with a cold start and no run duration limitation of 10 minutes! However, the pricing is a bit different. Where you would pay for the run duration and number of executions in the Consumption plan, with the Premium plan for Azure Functions, you will pay for the allocated vCPU, memory, and its usage. You would select this plan if you run functions continuously (or very often), or need any of the aforementioned features.
- **Dedicated plan**: This is exactly the same as an Azure App Service running in an App Service plan. In fact, a Dedicated plan means you will provision a regular App Service plan, which will then host your Azure Functions. It is ideal for long-running scenarios, but kind of defeats the point of serverless.

Now that we know what plans you can select to run Azure Functions on, let's take a look at supported operating systems and languages. To support a language, Azure Functions needs to support the *runtime* (such as .NET for C# and Node.js for JavaScript). You can run most on either the Windows or Linux operating system. Just like Azure App Service, the operating system is determined by the plan. Running a plan on Linux means all Azure Functions in that plan will run on Linux. There is more you can do in Linux Azure Functions, but more on that later.

A complete overview of supported runtimes and languages can be found here: `https://docs.microsoft.com/en-us/azure/azure-functions/supported-languages`.

Important note

There is more to the technology of Azure Functions than described in this chapter. As we will focus on running containers on Azure Functions, we will focus on explaining the concepts and components that you need to be aware of when venturing through the world of containers on Azure Functions.

Let's take a look at a use case.

Your company provides an e-commerce solution to the market. Your frontend runs on Azure App Service and you extend additional logic for order processing and user management to Azure Functions. You do this by having an Azure Service Bus in place that accepts messages from the frontend solution. Azure Functions is triggered when a message is available. Azure Functions performs the processing and reports back to the Azure Service Bus, avoiding any state stored in Azure Functions itself.

Your product manager informs you that they have purchased a licensed code to help them process billing statements. This code is an extension to the existing e-commerce platform, and it is preferred to run these in Azure Functions alongside the other logic. However, the code is written in a language with dependencies that are not natively available on Azure Functions. Despite your objections, it still needs to run, ideally in Azure Functions.

If we look at this use case, we see two requirements:

- The logic needs to run in Azure Functions.
- The code is written in a language not supported by Azure Functions yet.

And then, there's a problem. Perl is not a supported language on Azure Functions. But, what if we can package the code, the dependencies, and the runtime? What if we could run it in a container? Excellent idea!

If we dig a little further and take into account the concepts of cold start and Azure Functions plans, then we can see there is another challenge. In *Chapter 1*, we explained how containers work. There is an image that is pulled (downloaded) from a registry and a container is started based on that image. The bigger the container image, the longer it takes.

Back to Azure Functions in the Consumption plan. We stated earlier that we have to deal with a *cold start*. That means, upon calling Azure Functions, the dependencies are downloaded. After the Azure function is done processing again, it will go *cold* and deallocate, and all state will be gone. If we were to run a container in Azure Functions, that would require downloading the image each time we perform a cold start. That sounds like a problem. And it is. Luckily, Microsoft thought of that. Remember when we discussed the Premium and Dedicated plans? Exactly! They have the *warmed* workers, and resources are no longer deallocated.

Luckily, containers for Azure Functions are not supported on the Consumption plan because, well... that just wouldn't work with the downloading of the images, the cold start, and the startup times. Problem solved! We need a Premium or Dedicated plan. It's time to solve that use case and see whether we can help that product manager out.

Deploying containers to Azure Functions

Now that we understand the concepts of Azure Functions, let's see whether we can get a container up and running. First, we need a Premium or Dedicated plan to host Azure Functions on. That plan needs to run in a resource group. Let's create that first:

```
az group create `
--name "rg-functions-containers" `
--location "west europe"
```

> **Important note**
>
> Just like in the previous chapters, we used Azure Cloud Shell through the Azure portal to execute the Azure CLI commands.

Let's verify whether the creation of the resource group was successful:

```
az group show --name "rg-functions-containers"
```

You should have a similar result as shown in *Figure 3.1*:

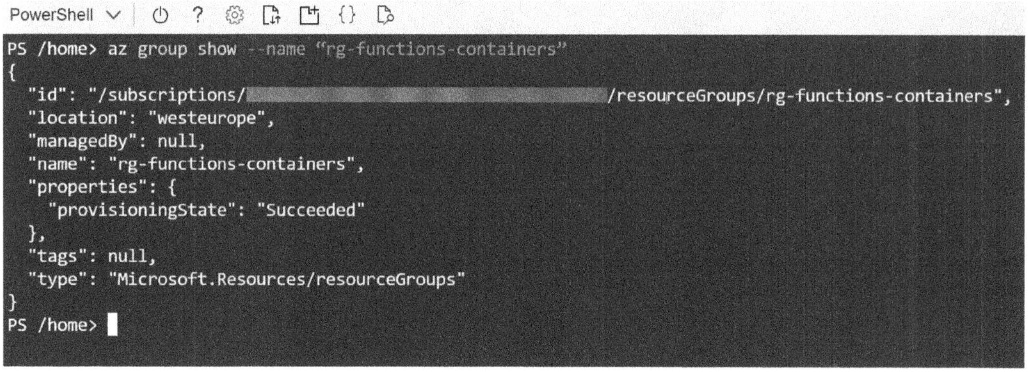
PowerShell
PS /home> az group show --name "rg-functions-containers"
{
"id": "/subscriptions/[redacted]/resourceGroups/rg-functions-containers",
"location": "westeurope",
"managedBy": null,
"name": "rg-functions-containers",
"properties": {
"provisioningState": "Succeeded"
},
"tags": null,
"type": "Microsoft.Resources/resourceGroups"
}
PS /home>

Figure 3.1 – Output of the az group show command

Now that we have the resource group, we can start deploying actual resources. For Azure Functions to work, we need storage. Luckily, Microsoft Azure provides us with storage accounts just for that:

```
az storage account create `
--name azfuncstor3421 `
--location "west europe" `
--resource-group rg-functions-containers `
 --sku Standard_LRS
```

> **Important note**
>
> Storage account names are required to be unique. For this, we added a random number to our storage account name. Of course, your naming convention may be different.

This will generate quite some output. Generally, a lot of output means that the deployment succeeded, but to verify, we can look for the `provisioningState` property with a value of `Succeeded`. This implies that the storage account is now operational. You can find the property just at the end of the output and it should actually be visible straight away:

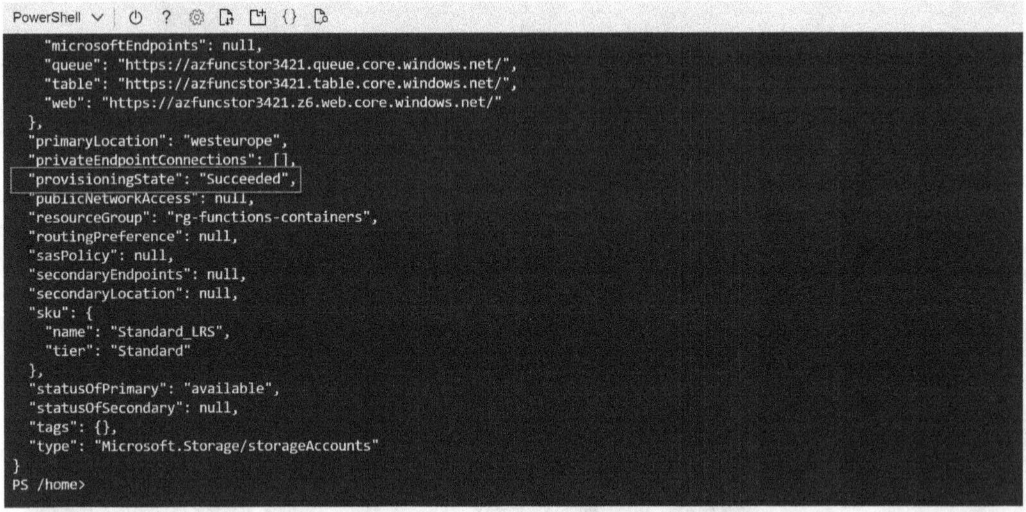
PowerShell
"microsoftEndpoints": null,
"queue": "https://azfuncstor3421.queue.core.windows.net/",
"table": "https://azfuncstor3421.table.core.windows.net/",
"web": "https://azfuncstor3421.z6.web.core.windows.net/"
},
"primaryLocation": "westeurope",
"privateEndpointConnections": [],
"provisioningState": "Succeeded",
"publicNetworkAccess": null,
"resourceGroup": "rg-functions-containers",
"routingPreference": null,
"sasPolicy": null,
"secondaryEndpoints": null,
"secondaryLocation": null,
"sku": {
"name": "Standard_LRS",
"tier": "Standard"
},
"statusOfPrimary": "available",
"statusOfSecondary": null,
"tags": {},
"type": "Microsoft.Storage/storageAccounts"
}
PS /home>

Figure 3.2 – Output of the storage account creation

Now that we have that, we are going to create the Premium plan that we discussed so many times already:

```
az functionapp plan create `
--name "containerplan" `
--location "west europe" `
```

```
--resource-group rg-functions-containers `
--number-of-workers 1 `
--sku EP1 `
--is-linux
```

Let's break this down. Once again, we will pass a name, location, and the resource group. We then define the number of workers. The number of workers defines the count of instances you will have available when executing an Azure function. The SKU determines the type of worker and its capacity (vCPU and memory). For now, we are going with one worker and using the `EP1` SKU. Additionally, we are selecting Linux as the operating system.

Again, this command results in quite some output but looking for that `provisioningState` property with a value of `Succeeded` will imply that our deployment was successful.

> **Important note**
> Azure Functions for containers is currently only supported for Linux plans.

Now, it is time to bring it all together. We are going to deploy the actual function app, connect it to the storage account we deployed previously, and host it in the Premium plan:

```
az functionapp create `
 --name containerinfunction01 `
--resource-group rg-functions-containers `
--storage-account azfuncstor3421 `
--plan containerplan `
--functions-version 3 `
--deployment-container-image-name nginx:nginx
```

> **Important note**
> It is currently not possible to deploy a custom image from your own Docker Hub directly from the command line. We will first use a basic nginx image to "populate" the function app with the option to configure a custom container image and use the Azure portal to set up the Docker Hub configuration.

Let's take a closer look at what we did here. We have created a function app named `containerinfunction01` and connected the `azfuncstor3421` storage account to the function app. This is required, as Azure Functions holds no state itself. We have attached the function app to the Premium plan (`containerplan`) and set the functions runtime version (this will be required in the feature). Last but not least, we used the `nginx:nginx` image as our initial deployment.

What we need to do now is check out how this all looks in the Azure portal and configure our container image that holds the `BillingStatements` solution (remember the use case?).

Let's open up the Azure portal (`https://portal.azure.com`) and navigate to our resource group, `rg-functions-containers`. There, we will find the resources we just created:

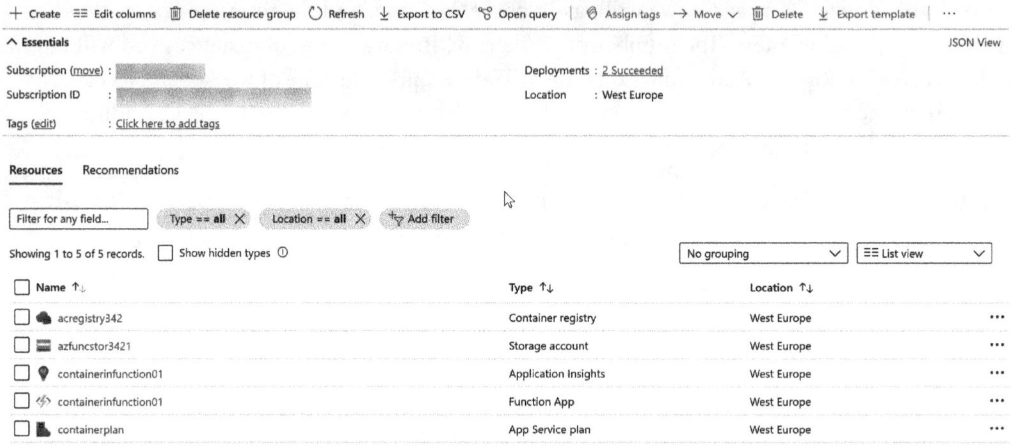

+ Create
Edit columns
Delete resource group
Refresh
Export to CSV
Open query
Assign tags
→ Move
Delete
Export template
Essentials
JSON View
Subscription (move) :
Subscription ID :
Tags (edit) : Click here to add tags
Deployments : 2 Succeeded
Location : West Europe
Resources
Recommendations
Filter for any field...
Type == all
Location == all
Add filter
Showing 1 to 5 of 5 records.
Show hidden types
No grouping
List view
Name
Type
Location
acregistry342
Container registry
West Europe
azfuncstor3421
Storage account
West Europe
containerinfunction01
Application Insights
West Europe
containerinfunction01
Function App
West Europe
containerplan
App Service plan
West Europe

Figure 3.3 – Resources created in previous examples

As we can see, all the resources are there. Now, it is time to configure our custom image and see whether we can get the `BillingStatements` solution up and running:

1. Click on the Azure function app (`containerfunction01`, in this example).
2. Navigate to **Deployment Center** in the left pane.

We are going to make a slight change to the existing configuration. As we have an image prepared for your convenience, we pushed this to Docker Hub so that it is publicly accessible.

We are going to make the following changes:

- Set **Repository Access** to **Public**.
- Set **Full Image Name and Tag** to `whaakman/container-demos:billingstatementsfunctionv01`.

We then save the configuration by clicking on **Save** on the top left of the window and wait for the function app to recycle its configuration.

Your configuration should look similar to the following screenshot:

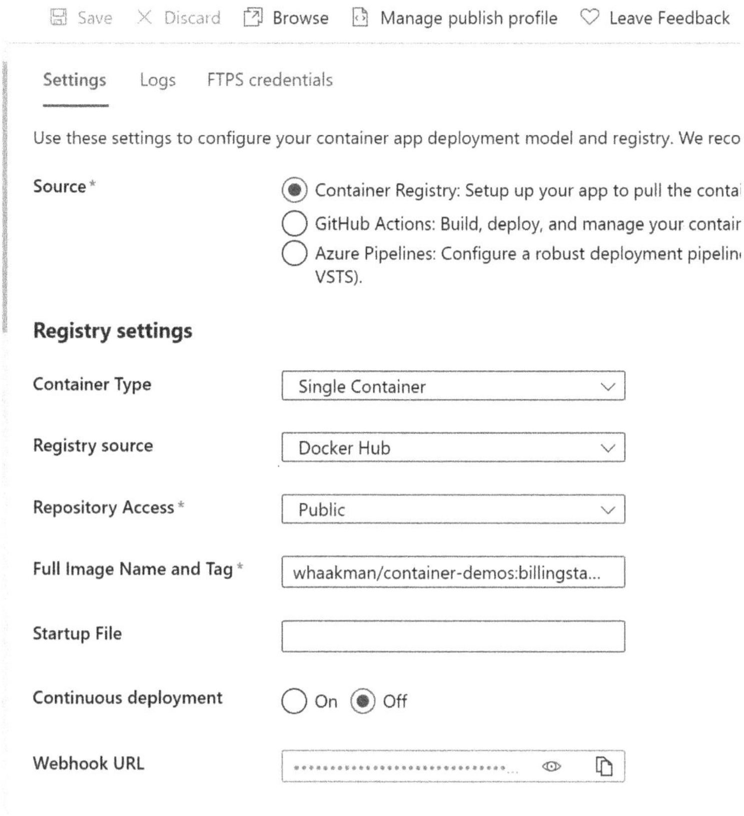
Save
Discard
Browse
Manage publish profile
Leave Feedback
Settings
Logs
FTPS credentials
Use these settings to configure your container app deployment model and registry. We reco
Source *
Container Registry: Setup up your app to pull the contai
GitHub Actions: Build, deploy, and manage your contair
Azure Pipelines: Configure a robust deployment pipelin
VSTS).
Registry settings
Container Type
Single Container
Registry source
Docker Hub
Repository Access *
Public
Full Image Name and Tag *
whaakman/container-demos:billingsta...
Startup File
Continuous deployment
On
Off
Webhook URL

Figure 3.4 – Deployment configuration for the Azure function app

That should be sufficient for us to run some tests. The container image we are using contains the .NET code for Azure Functions to service an endpoint called `BillingStatements`. Calling the endpoint should provide us with billing statement data.

As we know that our Azure function app is called `containerinfunction01`, we also know that Azure Functions publishes endpoints to the `/api/` path and that our function is called `BillingStatements`. This would result in the following URL: `https://containerinfunction01.azurewebsites.net/api/BillingStatements`.

> **Important note**
>
> Almost every deployment done through the Azure CLI or any other supported language will return a property with the endpoint URI. Additionally, you can request the URI through the Azure portal itself by browsing to the Azure function app.

Using PowerShell (or any tool you prefer), we can query the endpoint and see whether it returns the desired result:

```
Invoke-RestMethod https://containerinfunction01.azurewebsites.
net/api/BillingStatements
```

Let's see whether this works:

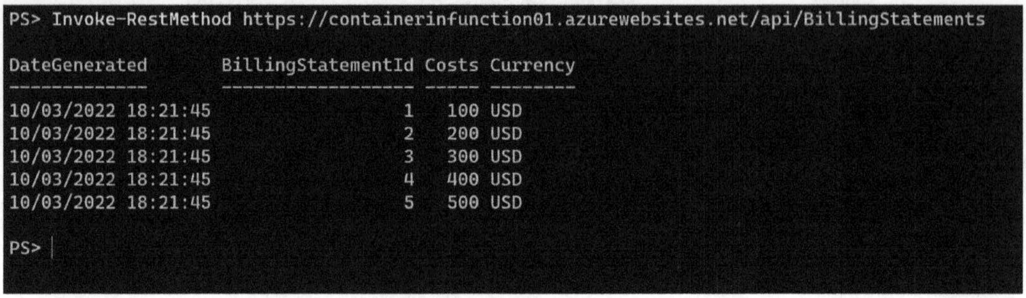
PS> Invoke-RestMethod https://containerinfunction01.azurewebsites.net/api/BillingStatements
DateGenerated BillingStatementId Costs Currency
10/03/2022 18:21:45 1 100 USD
10/03/2022 18:21:45 2 200 USD
10/03/2022 18:21:45 3 300 USD
10/03/2022 18:21:45 4 400 USD
10/03/2022 18:21:45 5 500 USD
PS>

Figure 3.5 – Results of the BillingStatements API

And, there you have it! Our container and API for the billing statements are up and running.

To ensure we do not end up with an invoice for unused resources at the end of the month, let's delete the resources we created during this chapter. We can do this by deleting the resource group using the following command:

```
az group delete --name rg-functions-containers
```

You have deployed both the nginx container image and a custom container image to Azure Functions. Now, let's dive a little deeper into Docker containers for Azure Functions.

Docker containers for Azure Functions

There are some particular quirks when it comes to building a container image for an Azure function. Where normally we can simply *containerize* the code we have built, for containers in Azure Functions apps, we have to follow some specific constructs. We have to implement the code as if we were writing an Azure function.

Now, that might sound like a very complicated thing to do, but in reality, it's not that hard. We can use tools such as the Azure Functions Core Tools to generate the base code for our function, the Dockerfile, and add our own custom code to the function to get started.

> **Important note**
>
> In the previous example, we built the container image for you to get started and test. To follow the code in this chapter, you will not have to do any of the following we are about to explain. However, if you are taking this to production, it's important to understand the basics.

To get started, we need the Azure Functions Core Tools, which you can download from `https://docs.microsoft.com/en-us/azure/azure-functions/functions-run-local`.

Let's do a test run. First, we need to initialize an Azure function and generate the Dockerfile. The Azure Functions Core Tools come with a `-docker` parameter to generate a Dockerfile out of the box. Additionally, we need to set the runtime we need. The reason we need to do this is that the Dockerfile will pull a base image containing the SDK and runtime that we select. If we need to execute `.Net` code, then we need the `.Net` runtime, and if we want to execute JavaScript, we need the Node.js runtime:

```
func init --worker-runtime dotnet -docker
```

This will scaffold a number of files but most importantly, a Dockerfile that contains the steps to build our container image.

Using the Azure Functions Core Tools, we can now generate a basic Azure function to correspond with the structure as defined in the Dockerfile:

```
func new --name BillingStatements --template "HTTP trigger"
--authlevel anonymous
```

This will generate an Azure function based on the `HTTP Trigger` template that is ready to be built and containerized. The generated code is the equivalent of a `Hello World` program but for Azure Functions. After running the previous commands, you should end up with a similar directory and file structure as the following screenshot:

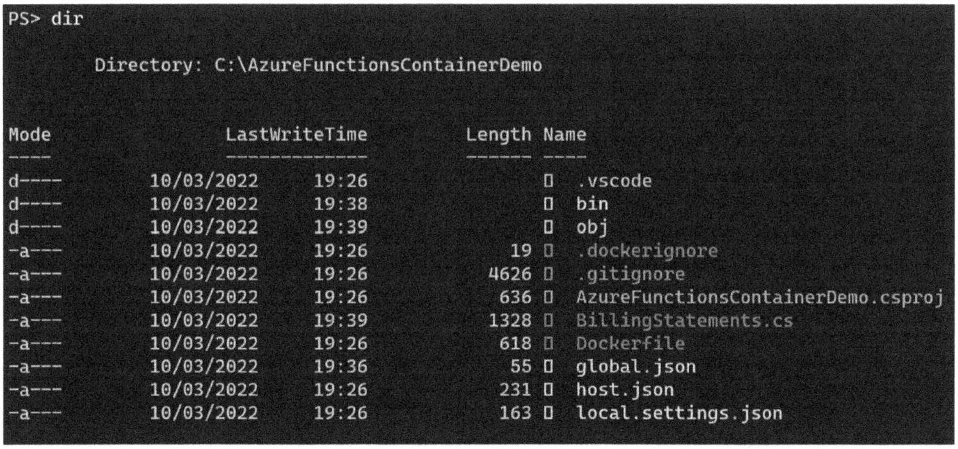
PS> dir
Directory: C:\AzureFunctionsContainerDemo
Mode LastWriteTime Length Name
d---- 10/03/2022 19:26 .vscode
d---- 10/03/2022 19:38 bin
d---- 10/03/2022 19:39 obj
-a--- 10/03/2022 19:26 19 .dockerignore
-a--- 10/03/2022 19:26 4626 .gitignore
-a--- 10/03/2022 19:26 636 AzureFunctionsContainerDemo.csproj
-a--- 10/03/2022 19:39 1328 BillingStatements.cs
-a--- 10/03/2022 19:26 618 Dockerfile
-a--- 10/03/2022 19:36 55 global.json
-a--- 10/03/2022 19:26 231 host.json
-a--- 10/03/2022 19:26 163 local.settings.json

Figure 3.6 – Directory and file structure of generated Azure function with Dockerfile

That looks like a lot of time saved, and it is! Azure Functions Core Tools is part of fulfilling the promise of Azure Functions where you *focus on writing the business logic*.

As we can see, the commands generated some files, and for now, two of these are important:

- `BillingStatements.cs`
- `Dockerfile`

`BillingStatements.cs` contains the basic code to run an Azure function. It makes the function respond to a `GET` or `POST` request, as we deployed the `HTTP Trigger` template. We can add additional classes (CS files) with models, logic, and whatnot. As long as we make the results end up in `BillingStatements.cs`, we should be fine.

Then, we have `Dockerfile`, which contains the commands to successfully build a container for Azure function apps.

> **Important note**
> We will not dive into the actual (proof of concept) code we wrote for this sample. However, the code is available in the Git repositories as part of this book.

We will leave the default (working) code for now and will start building the container image. In this example, we will push the code to Docker Hub, but the process is similar to pushing your container images to a service like Azure Container Registry. We chose to use Docker Hub so that the samples are available to you to experiment with.

To build our Docker image, we need to install Docker Desktop, which can be downloaded from `https://www.docker.com/products/docker-desktop`. Additionally, to push the image, you will need a Docker Hub account. During the installation of Docker Desktop, please make sure you select *Linux* as the default platform.

Alright, let's build that Docker container for Azure function apps. Inside the Azure function app directory that contains the Dockerfile, execute the following command:

```
docker build -t whaakman/container-
demos:billingstatementsexample .
```

This command will create the Docker image and provide it with a tag. In this example, we tagged the Docker image with `whaakman/container-demos:billingstatementsexample`. To push this image to Docker Hub, do the following:

1. Connect to Docker Hub.
2. Push the image to the account of `whaakman` in the `container-demos` repository and apply the `billingstatementsexample` tag/identifier.

This will push the image to our public Docker Hub repository, ready for use in Azure function apps, as we just described in the previous paragraph with the `BillingStatements` solution. You can view your pushed images by navigating to Docker Hub. In the preceding example, that would be `https://hub.docker.com/repository/docker/whaakman/container-demos`.

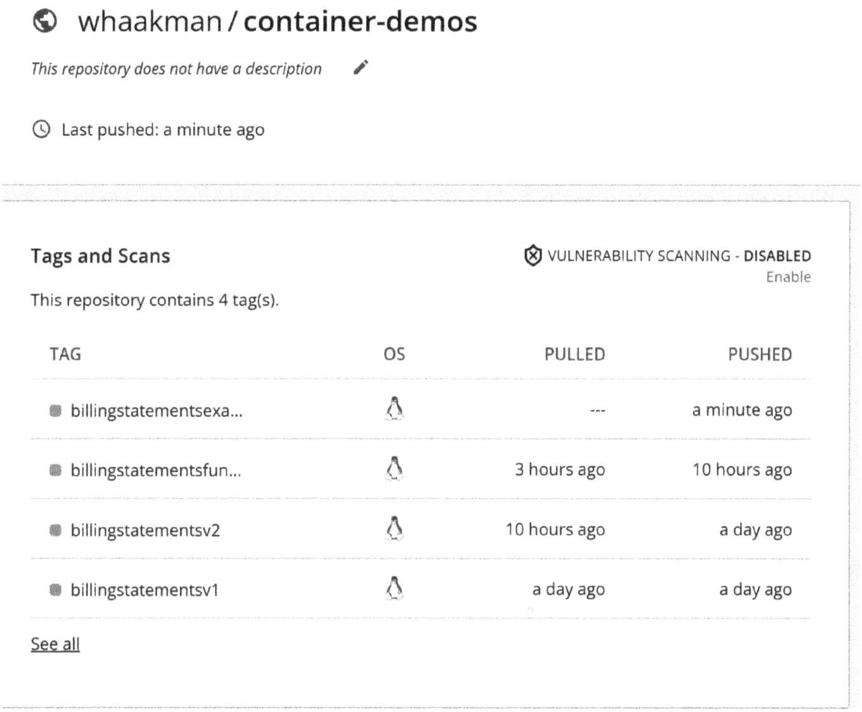
whaakman / container-demos
This repository does not have a description
Last pushed: a minute ago
Tags and Scans
VULNERABILITY SCANNING - DISABLED
Enable
This repository contains 4 tag(s).
TAG
OS
PULLED
PUSHED
billingstatementsexa...
---
a minute ago
billingstatementsfun...
3 hours ago
10 hours ago
billingstatementsv2
10 hours ago
a day ago
billingstatementsv1
a day ago
a day ago
See all

Figure 3.7 – Pushed images to Docker Hub

As you can see, the Docker images are pushed to Docker Hub and available for anyone to use.

In this section, you have learned how to generate the required code and standards to ready a container for use in Azure Functions. It is time to look at some pros and cons of deploying containers on Azure Functions.

Pros and cons of deploying containers to Azure Functions

As with any technology and as we have seen in *Chapter 2*, every technology has its pros and cons.

From this chapter, you might be aware it takes quite a few steps to containerize a solution for Azure function apps. We have also learned that Azure Functions are not the technology you would initially select as the primary choice for running a container. The use case works the other way around. You are using Azure Functions but now you need to run a solution with non-standard runtimes or dependencies. If we stick to that use case, it is an acceptable solution for running containers. Let's take a look at those pros and cons.

Pros

There are definitely a couple of reasons to use containers for Azure Functions.

It solves the problem of unsupported code and dependencies in Azure Functions

This is exactly the use case we described in the previous paragraphs. Your product manager confronts you with a piece of code that will not run in your standard Azure function. But, it so happens that the majority of your software architecture runs on Azure Functions. By containerizing the solution for Azure function apps, you are still able to run it within your existing standards.

But, that is not the only use case. What if you have experimental code in a runtime that is still running in preview? Maybe you can containerize it using the examples provided in this chapter and still run it on an Azure function until the runtime is officially supported.

Even though it can feel like a niche use case, once you get to it, the possibilities are endless.

It can run in your existing Azure Functions Premium plan

If you already have an existing Azure Functions Premium Plan or the dedicated App Service plan, you can deploy your container alongside your existing functions (provided this plan is using Linux as the target operating system). There is no need to buy an additional plan just for the containerized solution.

Speaking of that Premium plan, we could list the fact that containers are not supported on the Consumption plan. But, would you really want to wait for that cold start and a container image to be pulled from a registry before it runs? It is safe to say that Microsoft made a good choice focusing on Premium and Dedicated plans for containers on Azure Functions.

Integrations with Azure and beyond work out of the box

Because we are running an Azure function, we have access to integrations such as managed identities, Application Insights, Azure Container Registry, and CI/CD capabilities. We didn't specifically discuss the integrations, but they come out of the box. As the Azure Functions runtime will deal with those integrations, you do not have to worry about that within your code and container itself. We have already seen this with containers for Azure App Service, and sometimes this feels like a commodity. But, if you really think about it, it's quite exceptional.

Cons

With upsides come downsides, and some things in function apps for containers can be somewhat of a challenge. The main challenge is that you have to stick to the use case, but we already discussed that.

It's just Linux

It is currently not supported to run Windows containers on Azure function apps. It is currently unknown whether this feature will be introduced, but we do see Windows container support being introduced in other features. If we look at the world of Windows containers, it shouldn't come as a surprise. As we discussed in earlier chapters, the images are much larger, and using Windows containers still feels like you're using an operating system that the technology (containers) is not designed for.

However, with more modern languages such as .NET 5, .NET 6, and the upcoming .NET 7, cross-platform support is already present. So, does it really matter if Windows containers are not supported?

Have to follow Azure Functions standards

If you are being confronted with code that already exists and you have (or want) to use Azure Functions for whatever reason, you will have a small challenge. As we have seen in the *Docker containers for Azure Functions* section, you need to modify your code to Azure Functions standards to have the full support of the platform. That can be quite the task, but then again, if you are already using Azure Functions as a standard technology for your software architecture, you probably want that support and those integrations. Modifying the code accordingly may just be worth your while.

However, if the solution is short-term and you are planning to move to the actual native Azure Functions in the near future, we don't recommend making all those adjustments *just to have access to those integrations*. Use it as a dev/testing scenario and move on to more native Azure Functions technologies.

It's not the ideal solution to host containers

This is all about the use case. If you stick to it, you're fine. If you don't, you are in for a world of hurt. That might sound negative, but the Azure function apps are not designed to run containers. We are happy it is possible, but deploying the containers, making sure they are built correctly, and the debugging capabilities are limited and do not provide what we are used to when using other technologies.

Command-line tools don't support the full feature set

As we have determined in the examples earlier in this chapter, not everything can be done directly from the command line. To be fair, if you throw in a couple of additional commands, it will eventually work, but it would overcomplicate the process. Simply using the Azure portal for some of the configurations is the best way to do it.

That makes it complicated when you start thinking about pipelines and automation. Our best experiences come when using just Azure technologies. If we use Azure Container Registry to host our containers and configure a private registry, it works out of the box. But, even though we love Azure technologies, sometimes you are just bound to something else – Docker Hub, for example. Luckily, it's not a daunting process but the less we have to click, the better!

We can see that there are several pros and cons, but the cons for the most part are arguably not that bad. If you follow the standards as provided by Microsoft, things should all work out! On the other hand, Microsoft has provided us with the ability to overcome limitations in Azure Functions by introducing a container. Now, that is definitely a way to ensure the sky is the limit! The use case for introducing older dependencies or newer runtimes when they are not natively supported on Azure Functions is impeccable. And, the lack of support for Windows containers? We don't really mind.

Summary

In this chapter, we have discussed the capabilities that Azure function apps for containers bring. We learned that the use case is very specific and works the other way around; you will use Azure function apps for containers because you are unable to use the native runtimes or tooling that comes with Azure function app. It will never be your go-to platform for containers. Azure function apps for containers are there because you need an Azure function and your solution happens to be incompatible with Azure Functions.

The process for setting up a container for Azure Functions is pretty straightforward but might come with some trial and error to get it up and running (whoever said building a container image was easy?).

We have also learned that Azure Functions is *serverless* by nature. But also, to be truly serverless, we need to use the Consumption plan. Unfortunately, that does not work for containers, and with good reason.

In the next chapter, we will look at Azure Container Instances. If you want to talk about serverless containers, this is what you are looking for!

4

Azure Container Instances for Serverless Containers

It was July 2017; serverless technologies were upcoming, and with the ever-growing popularity of containers, Microsoft couldn't stay behind and released **Azure Container Instances** (**ACI**). Its promise is simplicity and no infrastructure management.

Over the years, the popularity of ACI has increased, and a lot has changed since its initial release date. ACI has become a valuable addition to the world of serverless and event-driven architectures. By integrating with numerous Microsoft Azure technologies such as Logic Apps, **Azure Kubernetes Service** (**AKS**), and Storage to name a few, the use cases are almost endless.

We see **machine learning** (**ML**) solutions, batch processing, and APIs being deployed to ACI. The promise of simplicity resulted in a technology where we merely need a few commands to get things up and running.

Throughout this chapter, we will help you understand how ACI works, build a solution based on a use case, and see whether it brings what is promised to the table.

In this chapter, we will cover the following topics:

- Understanding ACI
- Deploying containers to ACI
- The pros and cons of running containers on ACI

Understanding ACI

As we all know, containers are quickly becoming the go-to way to build cloud-native applications. With this fast innovation, we need a way to run containers in the cloud quickly and with no steep learning curve. This is where ACI comes into play. It is the fastest and simplest way to run both Linux and Windows containers in Azure. ACI even guarantees that your container is as isolated as it would be running in a VM. Unlike running a container in a VM, with ACI, you don't have to manage any underlying operating system, hardware, or patching that comes with running servers.

As mentioned in this chapter's introduction, ACI has numerous integrations with Microsoft Azure, but it has another trick up its sleeve: it can also be integrated with the Docker CLI. This is perfect for developers whose machines might not be up to the task of running multiple containers. By setting the Docker context to ACI via the command line, you automatically create an ACI when using the `Docker run` command. When you remove the running container, it automatically removes the newly created ACI. How cool is that?

Let's take a look at what makes ACI tick and what components it consists of, starting with container groups.

Container groups

Container groups are a top-level resource in ACI. Think of them like resource groups in Azure. But what exactly is a container group? Well, it is a collection of containers that get scheduled on the same host machine. Unlike an Azure resource group, every container inside a container group has the same life cycle and shares resources such as local network and storage. This is very similar to the concept of Pods in Kubernetes. We will dive further into Pods and Kubernetes in *Chapter 6*.

Let's have a look at an example container group:

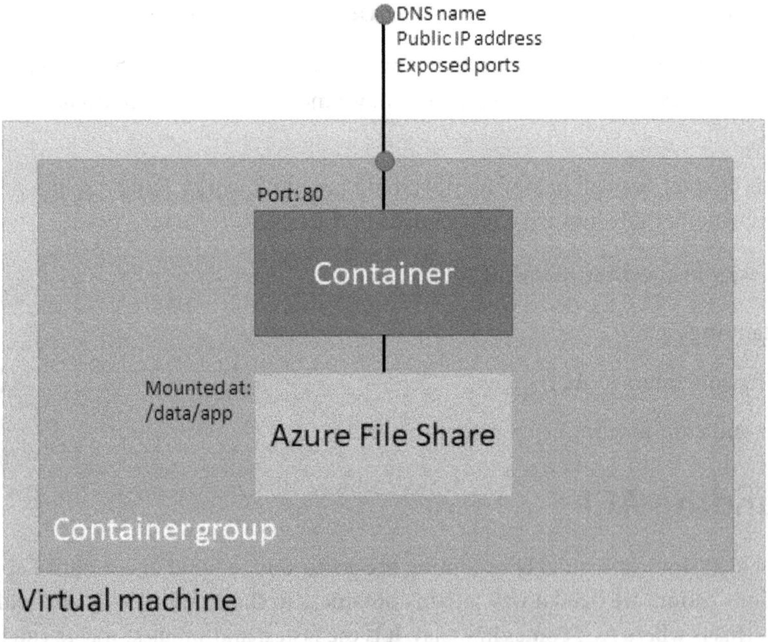
DNS name
Public IP address
Exposed ports
Port: 80
Container
Mounted at:
/data/app
Azure File Share
Container group
Virtual machine

Figure 4.1 – Overview of a container group

In this example, note the following:

- It is scheduled on one virtual machine (host machine).
- It has been automatically assigned a DNS name.
- It has a single public IP address and one exposed port.
- It contains one container.
- It has an Azure file share mounted to the `/data/app` directory locally.

Container groups come with some limitations concerning the maximum CPU and memory each container group can use. A container group can have a minimum of 1 CPU and 1 GB of memory. On average, they can have a maximum of 4 CPUs and 16 GB of memory for all containers running inside them. This can vary, depending on the Azure region. For an up-to-date list, go to `https://docs.microsoft.com/en-us/azure/container-instances/container-instances-region-availability`.

Say we have a container group that has four containers in it. Each container is configured to use a resource request of 1 CPU and 3 GB of memory. The container group will be allocated 4 CPUs and 12 GB of memory. If you needed to add another container to this container group, you would not be able to do so as it has hit the maximum of 4 CPUs. You would need to either adjust the current container's CPU resource requests or create a new container group.

If we only had one container that used a resource request of 1 CPU and 3 GB of memory, the container group would be allocated 1 CPU and 3 GB of memory. However, you also have the option to set a resource limit. This limit should be the same or greater than the resource request number. By using a resource limit, you are allowing your container to burst up to the limit you set. This may affect any other containers running in the container group.

Important note

The underlying infrastructure will use a small amount of the container group's resources, so it is always wise to ensure a small buffer when planning your resource allocations.

One last thing to mention about container groups is that they can share an external IP address with one or more ports on that IP address. They also share a DNS name to easily allow access to the containers running inside.

Talking about inside the container group, any containers running inside can talk to each other over the local host and any port. This is perfect for when your application needs multiple containers that communicate over a network.

When deploying an ACI, you have two options when it comes to networking. We'll look at these in more detail in the next section.

Networking

When it comes to networks, we can choose to deploy the containers *the standard* way, which will result in a public endpoint on which we can access our solution. However, ACI also supports integration with **Azure Virtual Network**, and with good reason. Many solutions that run on Azure are not just using one technology, such as ACI, but consist of several technologies. For example, parts of the processing are done in a container running on ACI, the frontend page is running on Azure App Service, and the database is running Azure SQL.

From a security and isolation perspective, there is always a question: do I want to use the public endpoints of these services, or do I want to implement a private network that only my infrastructure can access? Please note that we are talking about connectivity here; implementing a private network does not mean we can forget about the concepts concerning authentication.

ACI and many other services support Azure Virtual Network for that specific reason, for containing and isolating traffic.

There is a downside to this. When we isolate the traffic using Azure Virtual Network integration, we lose access to our public endpoints. There is no *and-and*; it is either private networking or a public endpoint.

We can overcome this by deploying other appliances in our virtual networks such as application gateways or even third-party appliances to route, scan, and secure traffic. If you think about it, using Azure Virtual Network integration features puts you in a completely different, more security-focused mindset.

Bursting from Azure Kubernetes Service

Now, here's a use case that makes ACI shine. ACI is more or less serverless. It runs when we need it. We still have to discuss a lot regarding AKS but let's go with the concept that your containers run on AKS, and you have allocated just enough resources for the average workload.

Your solution runs fine on AKS but it's an e-commerce solution that provides different customers with a web shop. Christmas is coming up and one of those customers decides to advertise an incredible discount for any purchase made on the website. You were not aware of this and the load on the platform skyrockets. You didn't plan for this, so AKS doesn't have the resources to meet the capacity requirements. Surely it can automatically scale, but it needs to be ready when the end users start visiting the web shop. You don't want to overprovision because 99% of the time, you don't even need the additional resources.

Alright; ACI to the rescue! By connecting ACI as a *virtual kubelet*, it essentially becomes part of the Azure Kubernetes cluster. Now, when the peak demand occurs, AKS will not automatically add nodes: it will spin up containers in Azure Containers. Once the demand lowers, the ACI containers will be stopped/deleted. Last but not least, we are only paying for the containers when we use them.

> **Important note**
>
> *kubelet* is the agent that is installed on Kubernetes nodes and deals with orchestration at the node level. You could say it takes orders from the Kubernetes control plane/API server.

Picturing a scenario is always helpful when learning about a new piece of technology. Once again, we are going to use a similar use case to what we used in *Chapter 3, Containers on Azure Functions? Sure You Can!*

> Your company provides an e-commerce solution to the market. Your solution is built on several different Microsoft Azure technologies.
>
> Your product manager informs you that they have purchased some licensed code to help them process billing statements. This code is an extension of the existing e-commerce platform and has several specific characteristics. The billing statement processing is not a continuous process. The processing happens when a customer requests their billing statements, which can happen at any random time of day. The processing is sort of a batch process with a peak load and needs to be available on demand. Deploying infrastructure that is continuously running would require considerable resource capacity, which makes it expensive as the billing statements are not being generated continuously.
>
> It is your job to get this up and running.

If we distill the requirements from the preceding use case, we end up with the following:

- The billing statement solution generates a peak load.
- The peak load can occur at any time of day when a customer requests their billing statements.
- We don't want infrastructure to run 24/7 unnecessarily.
- The solution needs to be available on demand.

You may have already guessed that containerizing the solution will be part of the solution. And that is correct! But we are going to use ACI as our primary technology.

Deploying containers in ACI

Now that we have familiarized ourselves with the concepts of ACI, we will use our `BillingStatements` solution once again and see whether we can get this up and running in an ACI. First, as with anything in Azure, we need a resource group:

```
az group create `
--name "rg-containers-aci" `
--location "west europe"
```

Important note

Just like in the previous chapters, we will be using Azure Cloud Shell through the Azure portal to execute the Azure CLI commands.

You should see an output similar to the following:

```
>> --location "west europe"
{
  "id": "/subscriptions/e378dee0-eef6-4335-bbd9-a3aa87496d91/resourceGroups/rg-containers-aci",
  "location": "westeurope",
  "managedBy": null,
  "name": "rg-containers-aci",
  "properties": {
    "provisioningState": "Succeeded"
  },
  "tags": null,
  "type": "Microsoft.Resources/resourceGroups"
}
PS /home> 
```

Figure 4.2 – Output from creating a resource group

Now that the resource group has been created, we have a place where our resources can live. The next step is to create a container group and deploy our billing statements application to it:

```
az container create `
--resource-group "rg-containers-aci" `
--name billingstatementscontainer `
--image whaakman/container-demos:billingstatementsv3 `
--dns-name-label billingstatements `
--ports 80
```

Let's go over some of the parameters in the preceding command. The first one, `-resource-group`, is an easy one. We just need to supply the resource group name that we used to create the resource group previously. Then, we have `--name`, where we need to supply a name for the container group. Next, we have `--image`. This is where we supply the path to the container image. In this example, we are using Docker Hub, which is the default container registry, so we don't have to supply it.

Now, it's time for `--dns-name-label parameter`. Here, enter a name you would like to use as the first part of the **fully qualified domain name** (**FQDN**). The FQDN in this example will be `billingstatements.westeurope.azurecontainer.io`. You will notice that the first part is the name you supplied, then the location where the container will be deployed, followed by `azurecontainer.io`. The last parameter is `--port`. You just need to supply the port number you want to access the container via. In this case, it's port `80` for HTTP.

> **Important note**
>
> The first time you deploy a container instance to a subscription, you might receive a message stating that **Microsoft.ContainerInstance Resource Provider is registering**. This is normal but may take a few minutes. You can learn more about resource providers at `https://docs.microsoft.com/azure/azure-resource-manager/management/resource-providers-and-types`.

After a short while, you should get an output similar to the following:

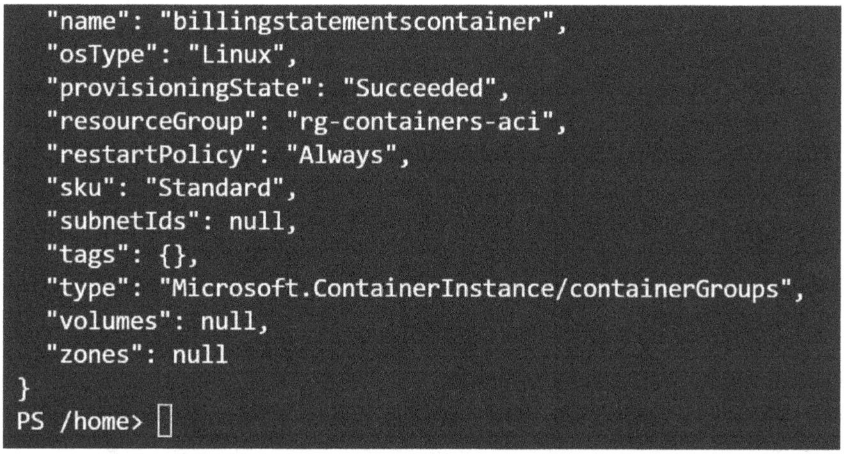
"name": "billingstatementscontainer",
"osType": "Linux",
"provisioningState": "Succeeded",
"resourceGroup": "rg-containers-aci",
"restartPolicy": "Always",
"sku": "Standard",
"subnetIds": null,
"tags": {},
"type": "Microsoft.ContainerInstance/containerGroups",
"volumes": null,
"zones": null
}
PS /home>

Figure 4.3 – Output of the az container create command

As ACI allows you to deploy multiple containers in a container group, the output may be very long, so you will probably not want to scroll through it to check for the FQDN and the current state of the container group. Luckily, we can use the `az container show` command to list this information:

```
az container show `
--resource-group "rg-containers-aci" `
--name billingstatementscontainer `
--query "{FQDN:ipAddress.
fqdn,ProvisioningState:provisioningState}" --out table
```

The output should look as follows:

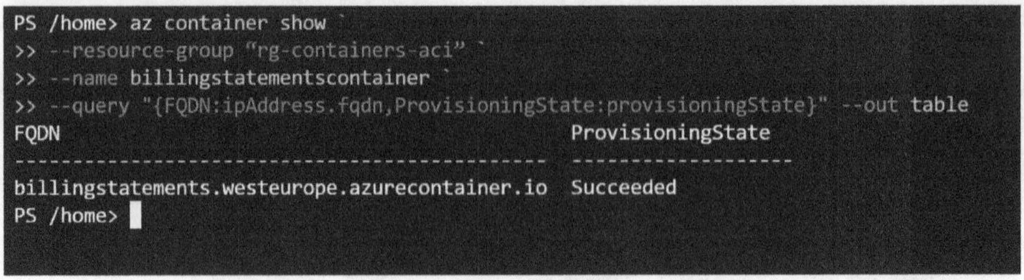
PS /home> az container show `
>> --resource-group "rg-containers-aci" `
>> --name billingstatementscontainer `
>> --query "{FQDN:ipAddress.fqdn,ProvisioningState:provisioningState}" --out table
FQDN ProvisioningState
billingstatements.westeurope.azurecontainer.io Succeeded
PS /home>

Figure 4.4 – Output of the az container show command

We now know the FQDN of the running container. The provisioning state shows the state of the container group. When it shows `Succeeded`, you can be confident that it has been deployed and the container has started. With the FQDN, you can use PowerShell to query the container. As we have seen in previous chapters, our example container providers a `/BillingStatement` endpoint. You can use the `Invoke-RestMethod` command in PowerShell to query the FQDN:

```
Invoke-RestMethod billingstatements.westeurope.azurecontainer.
io/BillingStatement
```

This should return a result similar to the following:

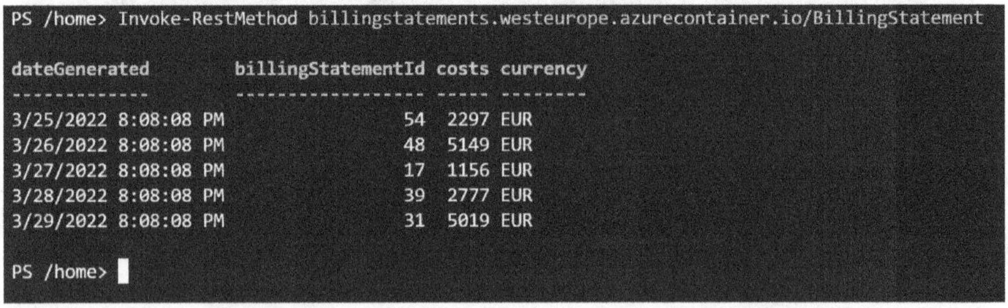
PS /home> Invoke-RestMethod billingstatements.westeurope.azurecontainer.io/BillingStatement
dateGenerated billingStatementId costs currency
3/25/2022 8:08:08 PM 54 2297 EUR
3/26/2022 8:08:08 PM 48 5149 EUR
3/27/2022 8:08:08 PM 17 1156 EUR
3/28/2022 8:08:08 PM 39 2777 EUR
3/29/2022 8:08:08 PM 31 5019 EUR
PS /home>

Figure 4.5 – Output of the Invoke-RestMethod command

You can also navigate to the same FQDN via your web browser. Here, you will be greeted with the same information but not in a user-friendly table:

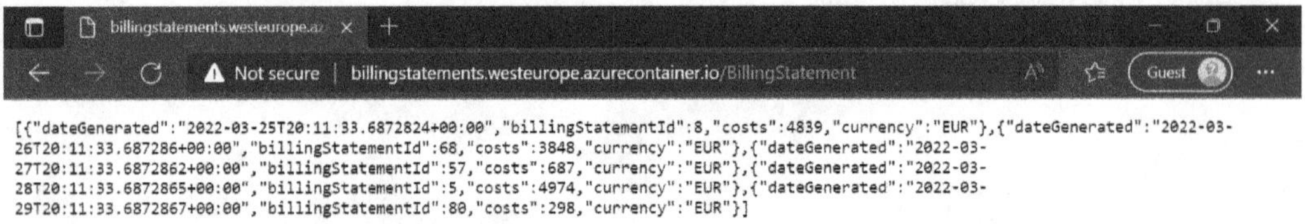
billingstatements.westeurope.az
Not secure
billingstatements.westeurope.azurecontainer.io/BillingStatement
Guest
[{"dateGenerated":"2022-03-25T20:11:33.6872824+00:00","billingStatementId":8,"costs":4839,"currency":"EUR"},{"dateGenerated":"2022-03-26T20:11:33.687286+00:00","billingStatementId":68,"costs":3848,"currency":"EUR"},{"dateGenerated":"2022-03-27T20:11:33.6872862+00:00","billingStatementId":57,"costs":687,"currency":"EUR"},{"dateGenerated":"2022-03-28T20:11:33.6872865+00:00","billingStatementId":5,"costs":4974,"currency":"EUR"},{"dateGenerated":"2022-03-29T20:11:33.6872867+00:00","billingStatementId":80,"costs":298,"currency":"EUR"}]

Figure 4.6 – Browsing to the container via a web browser

Now that we have our API up and running, let's take a look at the **Infrastructure as Code** (**IaC**) (here, YAML) that is required to achieve this.

ACI and YAML

Before we look at the pros and cons of ACI, there is an important fact we cannot leave undiscussed. In the world of containers, the YAML language is the standard for writing a Dockerfile that builds your container and configuration files for Kubernetes, Azure Container Apps, and several other technologies such as the Helm package manager. Up until now, we have discussed Azure App Service and Azure Functions, where, other than a Dockerfile, we did not have to deal with any YAML. In this chapter, we used the Azure CLI to deploy the Azure resources and a container to ACI. However, ACI also supports the use of YAML for creating multiple containers in container groups, as we discussed in the *Understanding ACI* section.

Why are we telling you this? In the second part of this book, we will dive more into the capabilities of each technology and see how they work in a real-world scenario, which is when a single container does not always cut it. After all, it is likely that you have a solution that runs on multiple containers (or are planning to build exactly that). Don't worry – we don't expect you to understand YAML from beginning to end just yet. When required, we will provide examples.

The reason why we are elaborating on the fact that ACI supports the use of YAML is that it is such an important language when you want to build scalable solutions using container technology. It is also an indication that ACI is built for running containers (hey, what's in a name?). This is opposed to Azure Functions and Azure App Service, which were originally developed to run code instead of containers.

We are using the Azure CLI for standardization throughout this book. It's readable, has the logic we are looking for, and does very well in a demo. If you looked closely, you might have seen the following parameter in the previous Azure CLI commands:

```
--image whaakman/container-demos:billingstatementsv3 `
```

This allows for exactly one container to be deployed. But what if our billing statements solution consists of a `BillingGenerator` solution that generates the bills, and a separate solution called `BillingStatements` to return the results to the user? That sounds like multiple containers! If we were to deploy such a solution, our best bet is to use YAML.

Where with a CLI such as Azure CLI we are telling ACI what to do step by step, YAML is descriptive. It is like telling the technology (in this case, ACI), "Hi there, this is the configuration I need; please get it done." Following is an example YAML file:

```
 # YAML Example
apiVersion: 2019-12-01
location: westeurope
```

```
name: billingcontainergroup
properties:
  containers:
  - name: billinggenerator
    properties:
      image: whaakman/container-demos:billinggeneratorv1
      resources:
        requests:
          cpu: 1
          memoryInGb: 1.5
      ports:
      - port: 80
      - port: 8080
  - name: billingstatements
    properties:
      image: whaakman/container-demos:billingstatementsv3
      resources:
        requests:
          cpu: 1
          memoryInGb: 1.5
  osType: Linux
  ipAddress:
    type: Public
    ports:
    - protocol: tcp
      port: 80
    - protocol: tcp
      port: 8080
type: Microsoft.ContainerInstance/containerGroups
```

The preceding YAML contains the configuration for two containers to be deployed to ACI. If we look closely, we will see that `properties` contains a property called `containers`. Exactly: plural. This means we can configure multiple containers to be deployed to the container group. Then, we define two containers, which requires us to create two container objects. They can be recognized by the `-name` properties at the beginning; we simply repeat the configuration of the first container to configure the second container.

For each container, we can define which ports it uses, what the resource requirements are, what environment variables need to be passed, or whether the container needs to be accessible to the outside world. We are not going into that level of detail just now but be aware that the YAML configuration usually brings a lot of capabilities.

Instead of using the list of CLI commands we used earlier, we will now only use the Azure CLI to deploy the YAML file to Azure. We will deploy to the existing resource group we created previously:

```
az container create `
--resource-group "rg-containers-aci" `
--file aci.yaml
```

> **Tip**
>
> As we are using Azure Cloud Shell throughout this book, you might be wondering, "How do I get that YAML into Azure Cloud Shell?" Azure Cloud Shell supports a Visual Studio Code interface. By typing `code aci.yaml`, a Visual Studio Code interface will be opened and you can paste the code into the interface. By clicking **save** or *Ctrl* + *S*, you can save the file and close the editor. Please note that you would normally do that in your home directory (`PS /home/username`) as that is where you have *write* access. Alternatively, you can get the file from this book's Git repository under the `chapter 04` folder.

Once again, lots of JSON output! We can search through the logs for an indication that the containers have been deployed. Alternatively, we can use the following command, which will return the deployed images:

```
az container show `
--resource-group rg-containers-aci `
--name billingcontainergroup `
--query containers[*].image -o json
```

The `query` part will search all container objects for the `image` property. The results should match the image names that we configured in the YAML file:

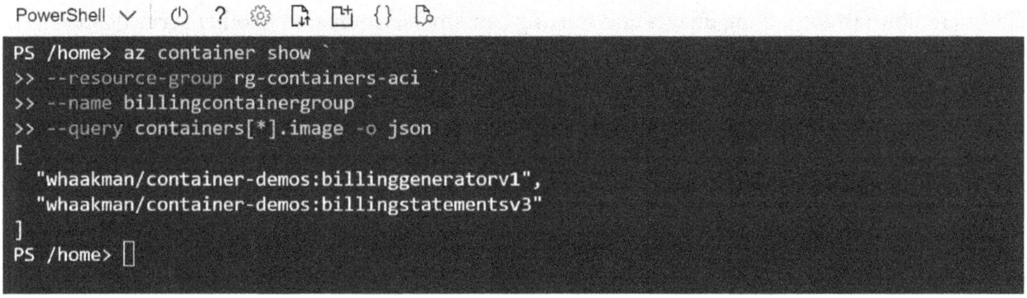
PowerShell
PS /home> az container show `
>> --resource-group rg-containers-aci `
>> --name billingcontainergroup `
>> --query containers[*].image -o json
[
"whaakman/container-demos:billinggeneratorv1",
"whaakman/container-demos:billingstatementsv3"
]
PS /home>

Figure 4.7 – Results of a multi-container deployment

Your result should look similar to what's shown in the preceding screenshot. With that, we have deployed two containers into a single container group. There is still a lot to be discussed on how your solution needs to behave. Specific characteristics of ACI and communication between containers can make it challenging to get your solution up and running without slight modifications. However, we are saving that for the second part of this book, where we will start scaling our use case from a start-up to an enterprise.

We don't want to end up with an invoice for unused resources at the end of this month, so let's delete the resources that were created in this chapter. We can do this by deleting the resource group using the following command:

```
az group delete --name "rg-containers-aci"
```

In this section, we created a container group and deployed a container to it. Now, let's go over some of the pros and cons of running containers on ACI.

The pros and cons of running containers on ACI

As you saw from this chapter, it is very quick and easy to start running a container or two in ACI. However, as with any technology, everything comes with pros and cons. In this section, we will go over some of the pros and cons of running containers in ACI.

Pros

Let's talk about the pros. What makes ACI good?

Consumption model

One of the biggest pros of ACI is that it is billed per second. As we can specify requests for CPU and memory, we are billed per GB of memory and per vCPU. This is perfect for when we need to run a container for a small duration or even just to burst from with AKS.

Fast startup times

ACI is incredibly fast for pulling images and starting containers. Compared to other services such as Azure Functions and Azure App Service, the performance is unmatched. This is because ACI is designed specifically for containers. Even if ACI might not be your target or final technology for running your containers, it can also be used as a resource to test whether your containers are working properly.

They work great in workflows

Previously, we mentioned that ACI has a consumption model; we said it has great startup times. So, what if we want to *consume* our container as part of a workflow? We can! **Azure Logic Apps** comes with a standard connector to run a container as part of a workflow. What? Really? Yes! That almost sounds like the way Azure Functions works. Turned off by default, business logic requires the container startups and shutdowns to be processed when we no longer need the container. This is a perfect use case for ACI.

Container groups

Having different container groups means we can isolate containers. As we are paying for consumption, it doesn't matter whether we deploy multiple container groups. But if we are in a scenario where we deploy multiple containers that do not need to communicate with each other and each focuses on servicing different business processes, why not separate them and use the isolation a container group brings? That's right! There's no reason not to do it.

Multiple containers

Because we have container groups, we can run multiple containers side by side and have them interact with each other. It is the first technology in this book we are discussing that has this capability. ACI is a safe bet if you're starting with a single container and potentially want to deploy additional ones over time.

Virtual network integration

Virtual network integration! While solutions such as Azure App Service and Azure Functions also provide this, for ACI, it is truly about integrating your container into the virtual network. It's not the most advanced feature, but if you want to control the flow of traffic to other services you are hosting in a virtual network and you want to prevent access from outside your environment, then this would be the feature you want – for every service. You could say virtual network integration is somewhat of a requirement for professional environments.

Cons

For most use cases, ACI is a very good solution. However, it does come with some limitations.

CPU and memory limits per container group

What if we have containers that need more resources than the limits of ACI? We would have to resort to a different technology. The hard limit of CPU and memory per container group is real. Luckily, for most solutions, you probably won't hit that limit and if you do, perhaps you can use a second container group. For more *monolithic* containers, these limits might be an issue, though it is understandable. ACI is built for speed, so by having containers that contain – let's put it mildly – less efficient code, you probably don't want to run them on ACI. Don't try to force it; choose the right technology.

If you want orchestration, start scripting

If your solution consists of multiple containers and the number of containers is growing, you might run into a scenario where you are doing a lot of things manually and you could use some help in the form of orchestration. Think about determining where the container is going to run (container group A or container group B), or whether you want to scale up or down. These are manual tasks.

If you want to automate these tasks within ACI, you need to start building some logic through scripts and then stitch together different Microsoft Azure technologies to get the desired result. Think about Azure Logic Apps, Azure Pipelines, and alerting based on telemetry to kick off such actions. The downside is you need to build this yourself. The upside is it is possible!

VNet integration and external access don't go well together

We can keep this short. If you use VNet integration, then VNet is what you use to communicate. Public endpoints are not available natively from ACI. This means you need to configure additional services to allow access from the outside world to your containers in ACI. It's more work but not always a bad thing. It does force you into a position where you need to think about what you are building, the networking, and then security. This may not be a huge downside but we would have liked it if we could have used both configurations together at the same time.

Windows Support

Windows Support is available, but it has a lot of catching up to do. For now, we recommend using it for experimental reasons or if you have to. Linux containers are a standard that has been around for many years, Windows containers are still relatively new to the game. It is admirable that there is support for Windows containers in ACI and we expect it to become better over time. But for now, ACI was designed with container standards in mind – standards built based on Linux.

We have now seen several pros and cons of ACI. And let's be fair – the pros are pretty good while the cons are not all that bad! Like with every technology, it is all about the use case. What we can see is that ACI is a technology that has been around for a while. It has features for natively integrating with Azure, including VNet integration, Logic App connectors, and AKS. The only serious downside we have seen is scalability and orchestration. If you are planning to run lots of containers and choose to use ACI, you are in for a challenge. ACI was not designed for that: it was designed for workflows that require incidental usage or simply for only a few containers. If you want to use it, you can, but we ask you to think it through. Other than that, we can safely say that it is a very good and mature technology to run your solution on.

Summary

In this chapter, you learned what ACI can do, what the technology is designed for, and how you can deploy it. You saw a very mature technology being able to run different types of workloads. We also introduced you to the YAML language, something that we will be using more often in the upcoming chapters in different shapes and forms.

ACI provides a lot and is a pretty big promise in general. But in the next chapter, we are going to dive into a more recently introduced technology: **Azure Container Apps**. What does Azure Container Apps bring to the table? Does it match up with what ACI is capable of? We will see in the next chapter!

5

Azure Container Apps for Serverless Kubernetes

At Microsoft Ignite in November of 2021, Microsoft announced a brand-new technology: **Azure Container Apps**, a fully serverless, application-centric hosting offering, meaning you do not have to worry about any underlying virtual machines, orchestrators, or other cloud infrastructure.

Throughout this chapter, we will help you understand how Azure Container Apps work, how to deploy a solution based on our well-known use case, and finally, see whether it brings to the table what is promised.

In this chapter, we're going to cover the following main topics:

- Understanding Azure Container Apps
- Deploying containers to Azure Container Apps
- Pros and cons of running containers on Azure Container Apps

Understanding Azure Container Apps

As mentioned in the introduction, Azure Container Apps is a new offering from Microsoft to easily run your container-based workloads in the cloud without having to worry about underlying infrastructure and orchestration. That does not mean that it does not use either. In fact, Azure Container Apps uses AKS—it is just managed by Microsoft and not yourself.

> **Important note**
> Azure Container Apps is currently, at the time of writing this book, still in public preview. As we expect plenty of development to take place, some examples may be outdated by the time you read this. The overall concept should still hold.

Azure Container Apps also makes use of other open source software such as **Kubernetes Event Driven Autoscaling** (**KEDA**) (`https://keda.sh`), **Distributed Application Runtime** (**Dapr**) (`https://dapr.io`), and Envoy (`https://www.envoyproxy.io/`).

KEDA is used to automatically scale your workloads if you choose to. It uses a concept of scalers, which include message queues, SQL queries, and even Azure DevOps pipelines. The list of scalers is ever-growing.

Dapr is used to help with communication between workloads, publisher/subscriber messaging systems such as Azure Service Bus, and more. We highly recommend you visit the Dapr website to find out its full potential. Incorporating Dapr into your solution will help you accelerate development and focus less on building logic for communication between components. Instead, the requirement for that communication comes out of the box with Dapr.

Envoy is used to enable Ingress for your workload. You can think of an Ingress as the router that contains the rules that allow traffic to your workload. Traffic from the outside world to your solution needs to be controlled. An Ingress does just that.

> **Important note**
>
> Ingress controllers and resources are a trivial part of scalable and secure container architectures. Up until now, we did not have to use any Ingress configuration as the **Platform as a Service** (**PaaS**) technologies in the previous chapters had their own technology in the background. There was no need for us to configure any Ingress. With Azure Container Apps, there is still no need for us to truly manage the Ingress. But the further we dive into the world of containers and the more our solution advances, we need control over that Ingress.

Think about enabling website access to your running workload. You are also able to create traffic splitting rules to help with A/B testing and Blue/Green deployments. Azure Container Apps will also create a **Fully Qualified Domain Name** (**FQDN**) and certificate to enable secure communication out of the box. Going forward, you will also be able to bring your own certificate and domain name.

Even though Azure Container Apps uses AKS in the background, it has some different concepts, which we will look into now.

Environments

Before you can deploy a container to Azure Container Apps, you need an environment. This environment is a security boundary for your running containers. Any containers running inside an environment are deployed into the same virtual network and write their logs to the same Azure Log Analytics workspace.

In *Figure 5.1*, you will see where environments sit with regard to Azure Container Apps:

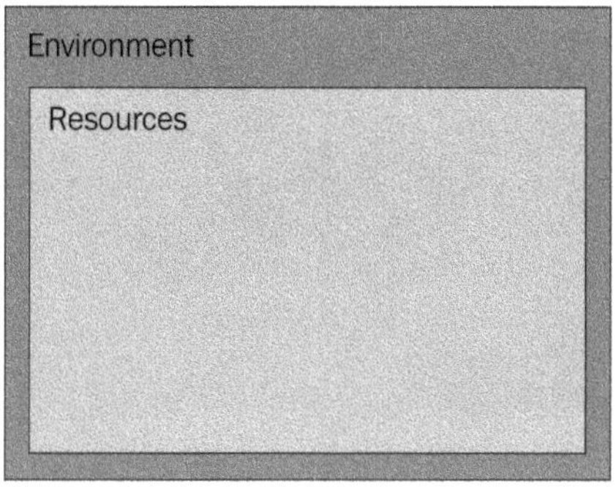
Environment
Resources

Figure 5.1 – Highlight of environments in Azure Container Apps

You can have multiple environments in your Azure subscription. In fact, if your applications do not need to talk to each other or they do not need to share the same resources, then it is probably better to keep them running in separate environments. It's good to note that you do not pay for environments, you only pay for individual container apps and their resource usage. One of the many benefits of serverless!

Some reasons you may want to deploy your workloads to the same environment are as follows:

- To share the same virtual network
- When applications need to communicate with each other
- When applications share the same configuration
- When applications need to log to the same Log Analytics workspace
- When they are part of the same solution

In Kubernetes terms, we could think of an environment as a *namespace* but one that is locked down with network and **role-based access control** (**RBAC**) policies.

Revisions

Revisions are immutable snapshots of your container app. They are a very important part of Azure Container Apps. Without them, you will be unable to deploy your container. Luckily, you don't have to create them as they are automatically created when you deploy your container app. *Figure 5.2* highlights where revisions play a part in the architecture of Azure Container Apps:

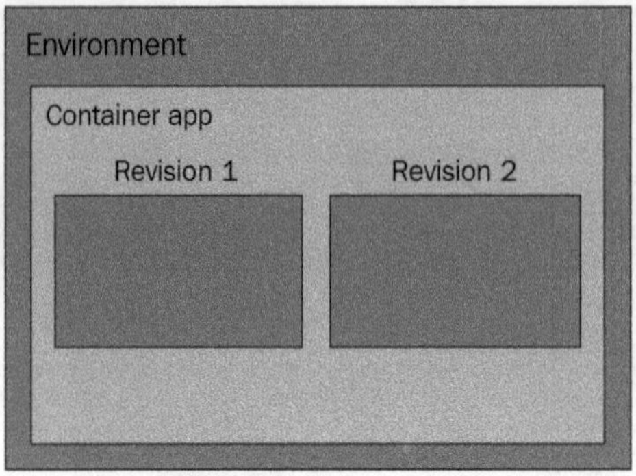
Environment
Container app
Revision 1
Revision 2

Figure 5.2 – Highlight of revisions in Azure Container Apps

When you update your container image or make changes to the configuration of your container app, a new revision is automatically created. Talking about changes in Azure Container Apps, there are two types and only one will create a new revision. Let's look at the two change types now.

Revision-scope changes

The following changes will automatically create a new revision:

- Changes to the container
- Add or update scaling rules (KEDA)
- Changes to Dapr configuration
- Any changes to the template section of the container configuration

Application-scope changes

The following changes will not create a new revision:

- Changes to traffic splitting rules (Envoy)
- Enabling or disabling Ingress (Envoy)
- Changes to secret values
- Any changes outside of the template section of the container configuration

In *Figure 5.3*, you will see a container app with two revisions and Ingress traffic splitting:

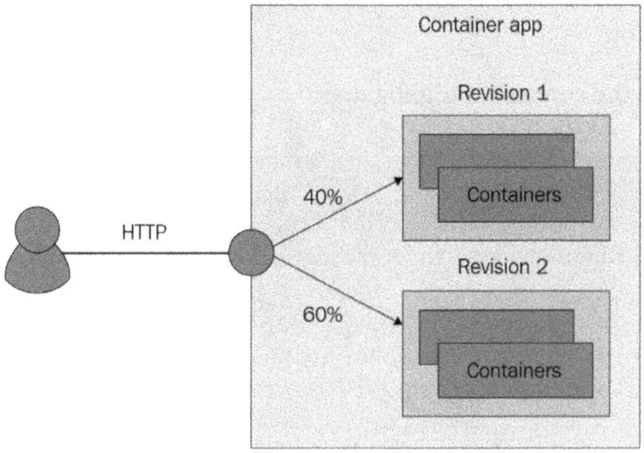
Container app
Revision 1
Containers
HTTP
40%
Revision 2
60%
Containers

Figure 5.3 – Ingress traffic splitting

In Kubernetes terms, I like to see revisions as a *ReplicaSet* but also as a *Service*. Just like revisions, ReplicaSets can be created automatically when you perform changes and updates on a Kubernetes *Deployment*. A Service in Kubernetes is used to help route traffic to your running container. With revisions, we get this automatically created for us when we enable Ingress.

Containers

Now, we are on to the most important part of Azure Container Apps, the actual container. Azure Container Apps can support any runtime, programming language, or development stack of your choice as long as it is using a Linux-based x86/64 (linux/arm64) container image. This image can come from any public or private container registry. Currently, at the time of writing this book, you can only use HTTP and HTTPS through ports `80` and `443` to access the container using the FQDN.

As you may have noticed from the previous sections, containers sit inside *Pods*. Pods come from Kubernetes and are the smallest deployable unit. A Pod can have one or more containers running inside it. This is the same with Azure Container Apps. In fact, when you use Dapr with Azure Container Apps, a sidecar container is also deployed. A sidecar container is a small container that sits beside your main container as a helper. For Dapr, this helper can be many things such as mTLS communication between different Pods.

Some reasons to run multiple containers together in a Pod are as follows:

- To use a sidecar such as Dapr
- Usage of shared resources such as disk space and local networking
- To scale together
- When both containers need to always run together (coupled)

Apart from the limitation of only Linux-based containers, you are also not able to use container images that require provided access. So if the container attempts to run a process that requires root access, you will get a runtime error.

In *Figure 5.4*, you will see the containers highlighted:

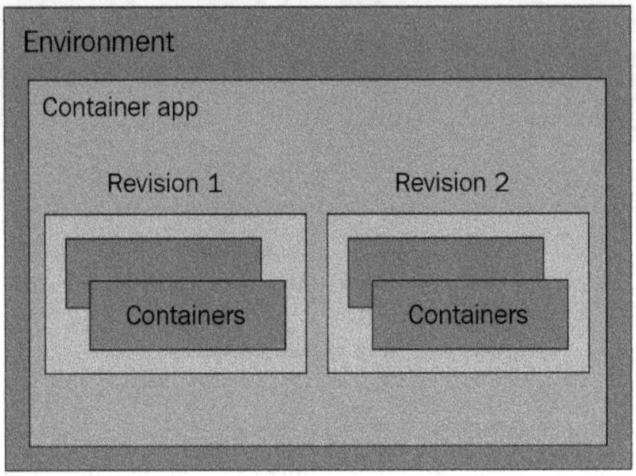
Environment
Container app
Revision 1
Revision 2
Containers
Containers

Figure 5.4 – Highlight of containers in Azure Container Apps

Containers, along with Pods, behave exactly the same as they do in Kubernetes. For example, if the container crashes, it automatically restarts. There is one small difference, however. You do not need to manually create the Pod in Azure Container Apps. This is done for you when you deploy your container.

At some point, we need to start deploying this. And like we did in the previous chapter, let's look at a use case to picture what we are trying to achieve.

> Your company provides an e-commerce solution to the market. Your solution is built on a number of different Microsoft Azure technologies.
>
> Your product manager informs you that they have purchased a licensed code to help them process billing statements. This code is an extension of the existing e-commerce platform and has multiple specific characteristics. The billing statement processing is not a continuous process. The processing happens when a customer requests their billing statements, which can happen at any random time of day. The processing is sort of a batch process with a peak load and needs to be available on demand. Deploying infrastructure to be continuously running would require considerable resource capacity, which makes it expensive as the billing statements are not being generated continuously. Additionally, we want to be able to control the isolation and communication between containers that might be added in the future, and the concept of environments in Azure Container Apps really appeals to you!
>
> It is your job to get this up and running.

We know it is still in preview but the features and level of control that Azure Container Apps provide make us decide that this is the technology we want to use.

Deploying containers to Azure Container Apps

Time to see whether we can help our product manager out. Can we get our billing statements API up and running in Azure Container Apps?

As this is a preview feature, the commands we need are not yet included in the Azure CLI tools. For that, we need to add the `containerapp` extension. We can do that using the following command:

```
az extension add -name containerapp
```

> **Important note**
>
> If you are using Cloud Shell through the Azure portal, it is not necessarily required to add the `az extension` up front. If it is not installed, Cloud Shell will prompt you and ask whether you want to install the required extension when you try to deploy a container app for the first time.

Before we can actually deploy a new Azure Container Apps app, we need to register the *Resource Provider* in our subscription. Normally, we are not required to do this, but it is common with preview features. We are simply telling Microsoft Azure "*Well hello there, we'd like to deploy Azure Container Apps, can you make sure that the required components to do that are available in my subscription?*". We can do that by executing the following command:

```
az provider register -namespace Microsoft.App
```

The prompt will return a message stating that the registration is ongoing and might take a while. This usually does not take more than a couple of minutes. By running the following command, you can verify whether the registration is successful:

```
az provider show -namespace Microsoft.App
```

As we have seen many times with the Azure CLI commands, we will be presented with a lot of JSON! Look for the `registrationState` property with a value of `Registered` to confirm that the Resource Provider has been registered:

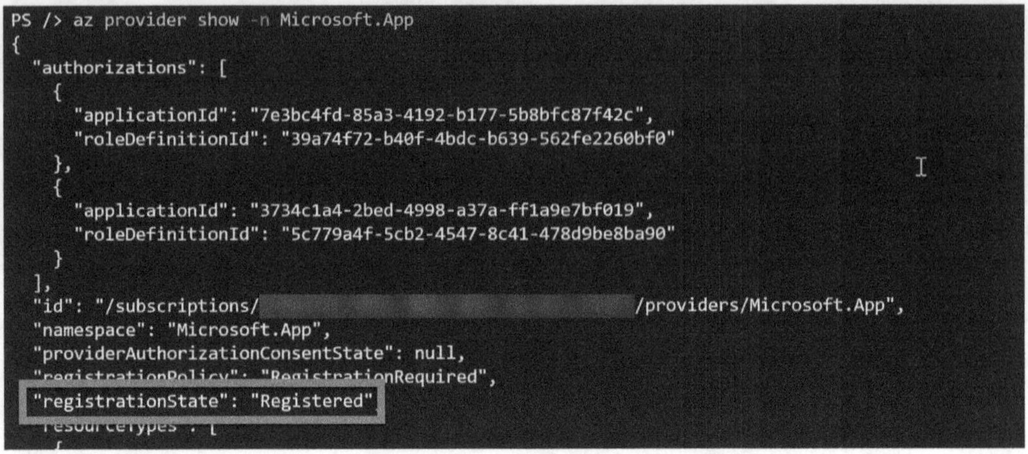

```
PS /> az provider show -n Microsoft.App
{
  "authorizations": [
    {
      "applicationId": "7e3bc4fd-85a3-4192-b177-5b8bfc87f42c",
      "roleDefinitionId": "39a74f72-b40f-4bdc-b639-562fe2260bf0"
    },
    {
      "applicationId": "3734c1a4-2bed-4998-a37a-ff1a9e7bf019",
      "roleDefinitionId": "5c779a4f-5cb2-4547-8c41-478d9be8ba90"
    }
  ],
  "id": "/subscriptions/[redacted]/providers/Microsoft.App",
  "namespace": "Microsoft.App",
  "providerAuthorizationConsentState": null,
  "registrationPolicy": "RegistrationRequired",
  "registrationState": "Registered"
```

Figure 5.5 – Confirmation that the registration has completed

The output should be similar to the preceding figure. Additionally, if you scroll down, you will see which regions are included in the registration, which also means that these are the regions you can deploy Azure Container Apps in.

> **Important note**
> It is common now that all Azure regions are supported by preview technologies. However, when features are released and hit **general availability** (**GA**), they usually cover the majority of the Azure regions worldwide. It has been known for Microsoft to release a new product to only a few regions at a time. So, please be aware of this when planning what technology to use.

We have now made sure our tooling (Azure CLI) and Azure subscription are ready to deploy and run Azure Container Apps. We are going to do just that right now!

First, we need a place for our Azure Container Apps app to live in. We can execute the following command to deploy a resource group:

```
az group create `
  --name rg-containerapps `
  --location westeurope
```

Almost instantly, a resource group is created, and you should get a similar output as in *Figure 5.6*:

PS /home> az group create `
>> --name rg-containerapps `
>> --location westeurope
{
"id": "/subscriptions/ /resourceGroups/rg-containerapps",
"location": "westeurope",
"managedBy": null,
"name": "rg-containerapps",
"properties": {
"provisioningState": "Succeeded"
},
"tags": null,
"type": "Microsoft.Resources/resourceGroups"
}
PS /home>

Figure 5.6 – Resource group created

We can see that we have created a new resource group called `rg-containerapps` with the `westeurope` location in our subscription.

Now that we have our resource group, we can deploy our environment. We are going with an environment for our billing management container, which we used in previous chapters as an example. The following command will deploy the environment for our Container Apps app:

```
az containerapp env create `
  --name billingmanagementapi `
  --resource-group rg-containerapps `
  --location westeurope
```

This command takes some time to execute. We have seen consistent deployment times ranging between 30 to 60 seconds. As long as the command indicates that the deployment is running as depicted in *Figure 5.7*, we are good:

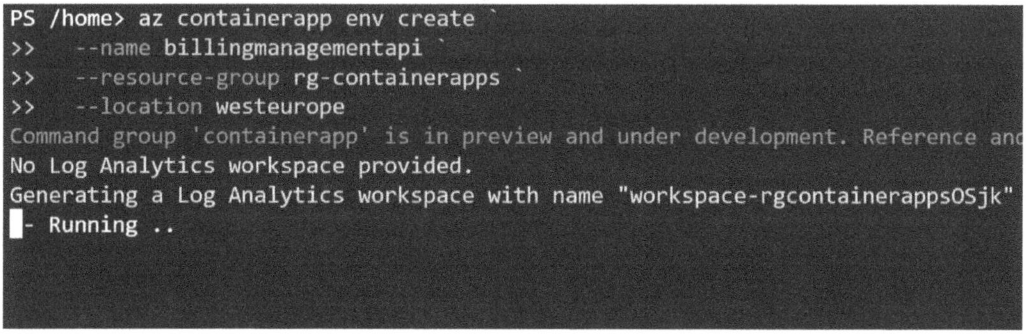
PS /home> az containerapp env create `
>> --name billingmanagementapi `
>> --resource-group rg-containerapps `
>> --location westeurope
Command group 'containerapp' is in preview and under development. Reference and
No Log Analytics workspace provided.
Generating a Log Analytics workspace with name "workspace-rgcontainerappsOSjk"
- Running ..

Figure 5.7 – Deploying the container app environment

A few things pop out here when executing the command to create the environment. We did not provide a Log Analytics workspace. Microsoft Azure will create one automatically. But, if you are deploying to a production scenario, you might want to add the parameter for defining which Log Analytics workspace you want to use. Why? Because the command we are using now will generate a random name.

If you already have an Azure Log Analytics workspace, you can use the following code to create an environment using it. If you executed the previous command to create an environment without a predefined workspace, you can skip running this command:

```
az containerapp env create `
  --name billingmanagementapi `
  --resource-group rg-containerapps `
  --logs-workspace-id myLogsWorkspaceID `
  --logs-workspace-key myLogsWorkspaceKey `
  --location westeurope
```

Now, we have everything we need to run our container. Let's create the container app within our environment! We will create a container app named `billingstatementscontainer` in the `billingmanagementapi` environment using the following command:

```
az containerapp create `
  --name billingstatementscontainer `
  --resource-group rg-containerapps `
  --environment billingmanagementapi `
  --image whaakman/container-demos:billingstatementsv3 `
  --target-port 80 `
  --ingress 'external'
```

As with the command to create the container app environment, this will take a couple of seconds to deploy. The time depends on your container image. As the container image is pulled from a registry, we have to wait for it to download. If we use the image we provided in our examples (the `billingstatements` container), it should only take a couple of seconds. But, be aware that deployment times may vary.

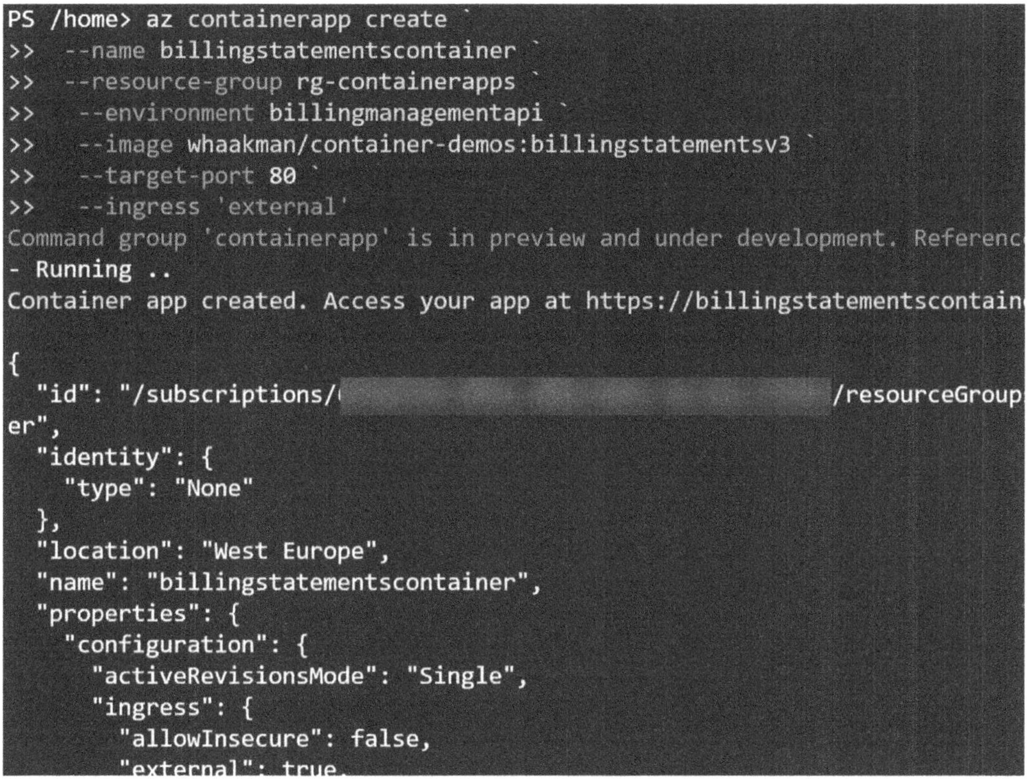
PS /home> az containerapp create `
>> --name billingstatementscontainer `
>> --resource-group rg-containerapps `
>> --environment billingmanagementapi `
>> --image whaakman/container-demos:billingstatementsv3 `
>> --target-port 80 `
>> --ingress 'external'
Command group 'containerapp' is in preview and under development. Referenc
- Running ..
Container app created. Access your app at https://billingstatementscontain
{
"id": "/subscriptions/ /resourceGroup
er",
"identity": {
"type": "None"
},
"location": "West Europe",
"name": "billingstatementscontainer",
"properties": {
"configuration": {
"activeRevisionsMode": "Single",
"ingress": {
"allowInsecure": false,

Figure 5.8 – Our container app has deployed successfully

Again, a lot of JSON output, but it starts with the address on which our container app is available. You can simply click on that or copy and paste it into your browser. However, if the address is not presented to you and you do not feel like going through the detailed JSON output, the following command will return the address of your solution:

```
az containerapp show `
  --resource-group rg-containerapps `
  --name billingstatementscontainer `
  --query properties.configuration.ingress.fqdn
```

The preceding command will return the address of our solution. As we have seen in previous examples, by knowing that address, we can attempt to connect to our API. Let's grab that URL.

```
PS /home> az containerapp show `
>>   --resource-group rg-containerapps `
>>   --name billingstatementscontainer `
>>   --query properties.configuration.ingress.fqdn
Command group 'containerapp' is in preview and under development. Reference and s
"billingstatementscontainer.calmdesert-2ad0a8c9.westeurope.azurecontainerapps.io"
PS /home>
```

Figure 5.9 – The endpoint for our container

Now, that is easier to read than going through the JSON result. Let us see whether we can query our `/BillingStatement` endpoint externally as we did in previous examples throughout this book:

```
Invoke-RestMethod https://billingstatementscontainer.
mangocoast-61009fce.westeurope.azurecontainerapps.io/
BillingStatement
```

We are using `Invoke-RestMethod` in PowerShell once again to achieve this. The result should look similar to the following:

```
PS> Invoke-RestMethod https://billingstatementscontainer.mangocoast-61009fce.
westeurope.azurecontainerapps.io/BillingStatement

dateGenerated       billingStatementId costs currency
-------------       ------------------ ----- --------
09/04/2022 08:38:19                 92  4210 EUR
10/04/2022 08:38:19                 46  3058 EUR
11/04/2022 08:38:19                 55  1310 EUR
12/04/2022 08:38:19                 61  5107 EUR
13/04/2022 08:38:19                 23   619 EUR

PS>
```

Figure 5.10 – Container containing the API is running

We can call that a success! The container returns the billing statements from our API as expected.

Because we configured `targetport` with a value of `80`, we told Azure Container Apps that our container listens on that specific port for traffic. Now, something very cool happens within Azure Container Apps. Once we provide the configuration as we did earlier using the `az containerapp` `create` command, Microsoft Azure will fill in the blanks. The standard is to expose your solution over a secure channel; when it comes to HTTP traffic, that would be port `443` secured with TLS. What Azure Container Apps does is detects that your solution is accessible over port `80` (because we told it to) and configures the Ingress to access traffic on port `443` (TLS), terminate the TLS connection, and send it to port `80` to your container.

Long story short, Azure Container Apps makes sure that your traffic to the Azure Container Apps endpoint is secure, even if you didn't think of it.

We can see what happened by looking at the visual representation of what we just deployed in the Azure portal:

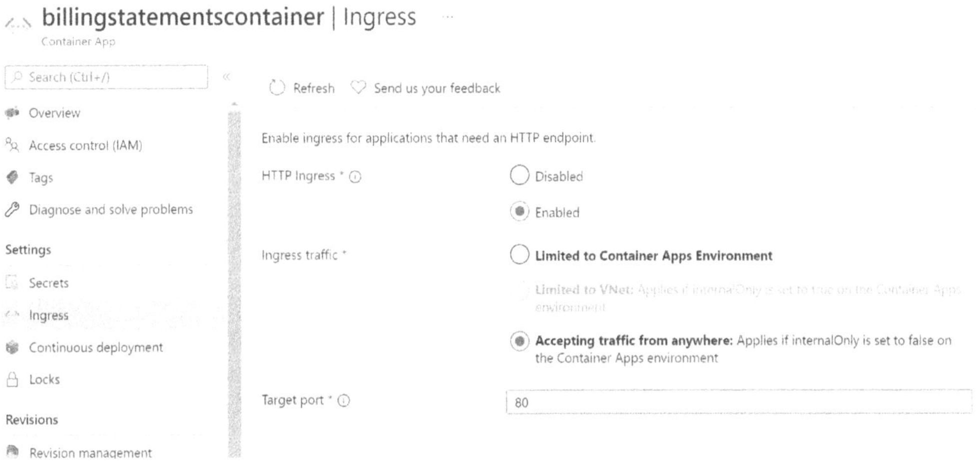
billingstatementscontainer | Ingress
Container App
Search (Ctrl+/)
Refresh
Send us your feedback
Overview
Access control (IAM)
Tags
Diagnose and solve problems
Settings
Secrets
Ingress
Continuous deployment
Locks
Revisions
Revision management
Enable ingress for applications that need an HTTP endpoint.
HTTP Ingress *
Disabled
Enabled
Ingress traffic *
Limited to Container Apps Environment
Limited to VNet:
Accepting traffic from anywhere: Applies if internalOnly is set to false on the Container Apps environment
Target port *
80

Figure 5.11 – TLS termination configured by default

But wait… where did the YAML code that we learned about in the previous chapter go? That is a good question! It went to deploying a single container to an environment. We can simply use the Azure CLI or the Azure portal. No need for additional Infrastructure as Code. However, if we were to deploy more complex environments, then yes, this would be a requirement. In the second part of this book, where we deep-dive into specific use cases and configure the infrastructure to support those use cases, we will explore the capabilities of running multiple containers across different environments and look into the capabilities of using Dapr and the Envoy proxy for communication.

We don't want to end up with an invoice for unused resources at the end of this month so let's delete the resources you created during this chapter. We can do this by deleting the resource group using the following command:

```
az group delete --name "rg-containers-aci"
```

In this section, you have created a container environment and deployed an Azure container app to it. Now, let's go over some pros and cons of running containers on Azure Container Apps.

Pros and cons of running containers on Azure Container Apps

We have seen that it is relatively easy to get started with a single container on Azure Container Apps. But of course, there are both pros and cons to using Azure Container Apps. Some of these will be explored in depth in *Part 2* of this book, and some we have already discussed throughout this chapter.

Pros

Let's start with some benefits that Azure Container Apps bring, and, surely, they are good.

Speed

Azure Container Apps is fast! Going from zero to an environment with a container running inside of it takes less than a couple of minutes and merely a few commands. If Azure Container Apps is not your production platform, it is definitely a platform to use for testing the functionality of your container in the Microsoft Azure cloud.

Microservices

Yes, microservices. The way Azure Container Apps is designed, the way we can isolate environments but also have multiple containers in a single environment, allows us to build sophisticated microservices architectures. With deep integrations with technologies such as Dapr and KEDA, Azure Container Apps is definitely the platform to build your microservices on if you don't need a fully-fledged AKS.

More control

With ACI, we did see some control over networking and inter-container communication. However, if you want more control over isolation, scaling, routing (Ingress), and better logging/monitoring, then Azure Container Apps is for you. It is friendly to the eyes and feels more intuitive.

KEDA and Dapr

If you say *cloud native*, then somewhere in your speech, KEDA and Dapr are mentioned. KEDA is the event-driven way to scale your solutions on Kubernetes and integrate with native Azure technologies. Dapr is its counterpart in the realm of development. It has grown to be quite popular, and for a developer not to worry about building complicated queueing mechanisms or session caching solutions, it's quite a time saver! Azure Container Apps provides these technologies out of the box.

Cons

There are not many cons that really stand out and, to be fair, Azure Container Apps is still in preview and some things may change. But, let's discuss what we have seen up until now.

Cold starts

They are serverless, so cold starts are a fact. Not that it has to pull your container image all the time but it's not on by default. Just like Azure Functions, you need to take this into account. As with anything, if your use case is different, choose a different technology.

The YAML is not YAML

Basically, we are talking about a representation of the YAML in JSON (ARM templates). That makes it harder to migrate from Azure Container Apps to a different platform (AKS, for example). It would require a rewrite of your configuration files to deploy containers. Where this is not a bad thing—especially if you are migrating to Kubernetes, where it is a good idea to understand what you are doing and build a Pod from the ground up—we would have loved to see a feature where we can export a Pod or deployment configuration for use in AKS. After all, Azure Container Apps is a layer on top of a managed AKS.

No control over ports for internal traffic

So, let's say we have two containers in an Azure Container Apps environment. One container provides us with the billing statements API, and one container serves as a backend with a Redis cache that is being leveraged by the API. Redis doesn't normally run on Port `80` or `443`. By default, it runs on port `6379`. At the time of writing, that's where the challenge comes in. We cannot control this in Azure Container Apps and were only able to find a solution using the standard ports. In order to resolve this, you would need to standardize the use of ports for your custom-built solutions to match what Azure Container Apps has to offer. This is a limitation that we might see disappear in the future (or at least we hope).

It is in preview

Azure Container Apps is still in preview. For the tech-savvy adventurists, that is perfectly fine! But while we are enthusiastic about the capabilities of Azure Container Apps, we can't use it in production yet. Maybe by the time this book is released, Azure Container Apps will hit GA. But for now, there are too many variables to take this into production just yet. However, for development and testing and to see whether your container is fit for the cloud, why not!

In our pros and cons, we can conclude that Azure Container Apps has a great future ahead. And rightly so. The cons we mentioned are either to be expected, such as cold starts, or are perhaps there because it is still in preview. We can't judge until it's finalized, but we are very positive about where this is going.

Summary

We ended the previous chapter with "*What does Azure Container Apps bring to the table and does it match up with what Azure Container Instances are capable of?*". To answer that question: we think it can!

Azure Container Apps is a technology designed to make more complex infrastructures such as Kubernetes easier. As we have mentioned, AKS is the technology powering this, but the complexity is abstracted away by providing a layer on top of it: Azure Container Apps.

We have seen Azure Container Apps holds great promise for the future and will become a go-to platform for developers to get started building containerized solutions. As Azure Container Apps comes with out-of-the-box configurations for KEDA, Dapr, and Envoy (Ingress), it demands best practices to be used in code. This is a good thing.

In our pros and cons, we have seen and can easily conclude that the pros are already very good and the cons might be taken away when Azure Container Apps hits GA.

Then there is the question, "*can I go beyond Azure Container Apps?*" Well, you definitely can. In the next chapter, we are going to take a look at AKS, see what it can provide, and learn about the technology that also powers Azure Container Apps.

6

Azure Kubernetes Service for Kubernetes in the Cloud

Initially, Microsoft started with **Azure Container Service** (**ACS**). ACS allows you to pick which container orchestrator you would like from Mesos, Docker Swarm, and Kubernetes. As time went along, it became clear that Kubernetes had won the container orchestrator wars and Microsoft decided to go all in on Kubernetes. On October 26, 2017, Azure Kubernetes Service went **Generally Available** (**GA**), and since then, the product has grown along with the Kubernetes community.

Throughout this chapter, we will help you understand how **Azure Kubernetes Services** (**AKS**) works, build a solution based on a use case, and see whether it brings to the table what it promises.

In this chapter, we're going to cover the following main topics:

- Understanding how AKS works
- Deploying containers to AKS
- The pros and cons of running containers on AKS

Understanding how AKS works

To start off, let's have a little recap about what Kubernetes is.

According to the official Kubernetes docs (`https://kubernetes.io/docs/concepts/overview/what-is-kubernetes/`):

Kubernetes is a portable, extensible, open source platform for managing containerized workloads and services, that facilitates both declarative configuration and automation. It has a large, rapidly growing ecosystem. Kubernetes services, support, and tools are widely available.

But what does this mean? It basically means Kubernetes is a platform to run your container workloads. It will do what it can to ensure that what is running on the Kubernetes cluster matches the configuration you have declared you want to run.

AKS uses the open source Kubernetes project and builds on top of it to ensure it can run inside the Microsoft Azure cloud.

An AKS cluster is made up of two parts – a control plane, which is used for the core Kubernetes Services, and nodes, which are used to run your workload and some Kubernetes system Pods, such as CoreDNS.

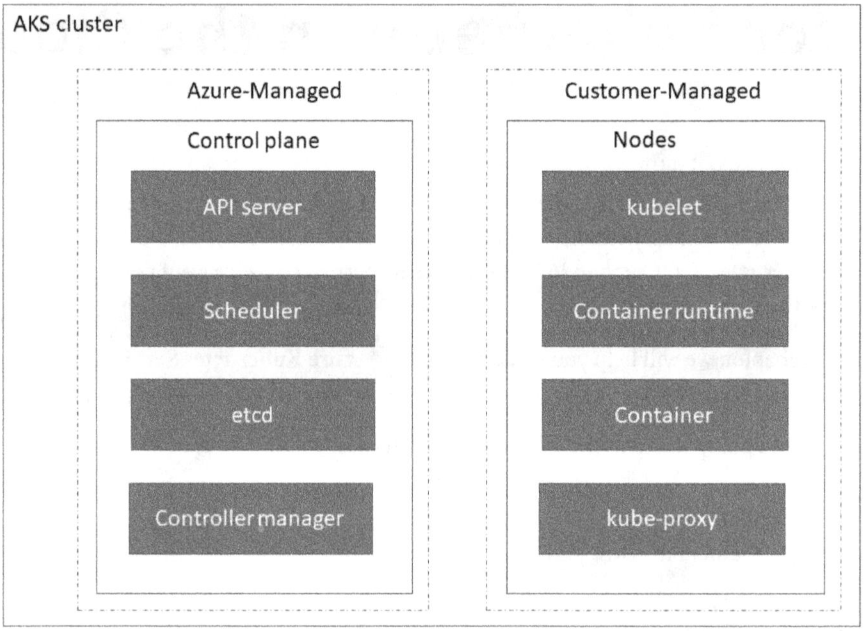
AKS cluster
Azure-Managed
Control plane
API server
Scheduler
etcd
Controller manager
Customer-Managed
Nodes
kubelet
Container runtime
Container
kube-proxy

Figure 6.1 – Azure-managed Kubernetes versus customer-managed

Let's go deeper into the two parts.

Control plane

When you are creating an AKS cluster, the creation of the control plane is handled for you automatically and is provided to you at no cost. Due to this, you do not actually have access to the control plane directly, only via the APIs, command-line tools, or the Azure portal.

> **Important note**
>
> You are still able to troubleshoot the control plane by exposing the control plane logs through Azure Monitor, but this will incur extra costs. This feature can be enabled through the Azure portal. You will find it under **Diagnostic settings** on the AKS blade.

When deploying an AKS cluster, you can pick a few options, one of them being location. Say you choose West Europe when deploying your AKS cluster; the control plane will then be created in that region. So, basically, the control plane in AKS is not a global Azure resource.

As mentioned earlier, the control plane is used for the core Kubernetes Services such as the following:

- **kube-apiserver**: This component provides the API that tools such as `kubectl` interact with.
- **etcd**: This is the central store where the state of your AKS cluster is stored. It is a highly available key-value store.
- **kube-scheduler**: This component determines what node can run the workloads and starts them.
- **kube-controller-manager**: This component, as the name suggests, is the manager that oversees smaller controllers. It's kind of like the captain of a ship.

Even though the control plane is free, it is single-tenanted, with each of the components in the preceding list dedicated to you and your AKS cluster.

Nodes and node pools

When we say nodes and node pools, what we really mean are Azure **virtual machines** (**VMs**) and **Azure Virtual Machine Scale Sets**. These VMs are where your workload and Kubernetes-supporting Services actually run. Each node will have the following components:

- **kubelet**: This Kubernetes agent listens to the orchestration requests from the kube-scheduler that runs on the control plane and then processes the requests.
- **kube-proxy**: This component handles the virtual networking on each node. It helps route traffic and manages the IP address for Services and Pods.
- **Container runtime**: This component allows the containers to run. It used to be Docker, but since version 1.19 for Linux and 1.20 for Windows, it has been swapped for containers.

You are able to use most Azure VM sizes when working with AKS, but it is good to note that the size of your node defines the amount of CPUs and memory your workloads have. By default, AKS nodes are created with premium SSDs, but you can also use ephemeral disks with AKS. Not all VMs support this, so ensure you pick the node size to meet your workload demand.

Nodes of the same size and configuration are grouped into a node pool. When deploying an AKS cluster, you will always have at least one node pool. This could contain just one node or a number you decided when deploying the AKS cluster. This first node pool is always the system node pool.

System node pool

As mentioned previously, the system node pool is always the first node pool created in an AKS cluster. This does not mean that you are unable to create a new one at a later date, but you do need to have at least one system node pool in your AKS cluster.

The system node pool is used to run system Kubernetes Pods such as `coreDNS` and `metrics-server` to name a few. By default, you are also able to run your own workloads on this node pool, but it is not recommended for production usage.

Some features and limitations of system node pools are as follows:

- They only support Linux.
- They can have just one node pool, but it is recommended to have a minimum of two nodes, or three if it is your only node pool.
- Each node needs to have at least two vCPUs and 4 GB of memory.
- They have to support a minimum of 30 Pods.
- Spot VMs (`https://docs.microsoft.com/en-us/azure/virtual-machines/spot-vms`) are not supported.
- You can have multiple system node pools.
- If you only have one node pool, it cannot be deleted.
- It can be converted to a user node pool if you have another system node pool.

As you can see, system node pools are very important and a much-needed resource. It is highly recommended to not run your workloads on a system node pool but rather run them on a user node pool.

User node pool

With user node pools, you have a few more options over system node pools. The main one is Windows support. That's right! You can create a Windows node pool to run your Windows container workloads. This user node pool will only run Windows containers and not Linux; they are not multi-OS hosts. Some more features and limitations are as follows:

- They can scale all the way down to 0 nodes.
- They can be created and deleted with no issues, unlike system node pools.
- You can use spot VMs.

- They can be converted to a system node pool if it is using Linux.
- You can have as many user node pools as Azure will let you.

As you can see from the preceding list, user node pools are rather cool. As you can have multiple pools each with a different node size, you can use Kubernetes features, such as taints and tolerations (`https://kubernetes.io/docs/concepts/scheduling-eviction/taint-and-toleration/`), to spread your workloads over different node pools to ensure you utilize resources to the max.

Now that you have an understanding of the node pool types, it's time to take a look at the Kubernetes objects. Let's start off with namespaces.

Namespaces

Namespaces are a way to logically group Kubernetes resources together, just like a resource group in Microsoft Azure. Out-of-the-box AKS comes with three different namespaces:

- **default**: When you deploy Kubernetes resources without specifying a namespace, this is where they will end up. For small environments, it is okay to use this.
- **kube-system**: This is where core resources live; most of the core resources you find in the system node pool are here.
- **kube-public**: This is not typically used but it is designed for resources to be visible across the whole cluster and viewed by any user.

You are able to set namespaces as a security boundary. This enables you to split a Kubernetes cluster into separate business groups, and only users assigned to that business group can access the resources inside the namespace.

Now that you know what a namespace is, it's time to take a look at what you can deploy into them. Let's start with the smallest deployable object in Kubernetes, Pods.

Pods

Pods are the smallest deployable unit in Kubernetes and AKS. Each Pod represents one instance of your application. You typically have one container per Pod, but in more advanced scenarios, you may need to run multiple containers per Pod, which is supported.

Pods are designed to be disposable. This means that at any moment, the Pod may be deleted, and your application will no longer be running. But that's okay, as Kubernetes has a fix for that called deployments.

Deployments

A Deployment is basically a collection of identical Pods. You can specify the amount of Pod *replicas* to create, and the Deployment will ensure that there are always that many Pods running. So, if one gets deleted, then the Deployment will start a new one to ensure your application stays working.

As you have the ability to ensure the specified number of replicas are running, it is highly recommended to use deployments for most of your workloads in AKS over a single Pod.

Services

In Kubernetes, each Pod gets its own IP address, and, as we know, Pods are designed to be ephemeral. This means that with every new Pod, or restart of a Pod, a new IP address will be used. This is no good if you have an application communicating with this Pod via its IP address and then it changes. You will have to update the configuration with the new IP address each time.

Then, with deployments, we can have multiple replicas of one Pod running, each with its own IP address. How do we load-balance over the Pods? That's where Services come into play. They are a dependable endpoint within Kubernetes that help to distribute traffic to Pods that matches a label. This endpoint has both an IP address and an internal DNS name for your cluster.

There are a few different Service types that you can use, depending on your application. The three main Service types are as follows:

- **ClusterIP**: This will give the Service an IP address internal to the cluster only.
- **NodePort**: This service type will expose the Service on each node's IP address but with a static port. It also creates a ClusterIP. You can connect externally to the cluster by using the node's IP and the static Service port that was created.
- **LoadBalancer**: This Service type will expose the Service externally via an Azure load balancer. When you choose this Service type, the other two Service types are automatically created for you too.

So, you could use a `LoadBalancer`-type service to expose your application to the world, but it does not come with TLS termination or complex routing. That's where Ingress objects come into play.

Ingress

In the previous chapters, you will have noticed one thing. The Ingress, the way to access the application, has been created for us automatically. That is not the case with AKS. You need something called an Ingress controller. This is a bit of software that runs inside your AKS cluster to, in a way, act as a router for your applications. This Ingress controller will use an Azure load balancer and assign a public IP address. Any requests to this public IP address will be routed to the Ingress controller Pods. The Pods will then check something called Ingress rules to see where to route the traffic.

Ingress rules are Kubernetes objects that are created via manifest files. Each file has a set of rules. These rules have information about the hosts they relate to (say, `billingexample.packt.com`), a list of paths (say, `/testpath`), and then which Kubernetes Service/Pods it needs to talk to.

Once the Ingress rules are deployed to the AKS cluster, the Ingress controller will load in the rules and start to serve the traffic.

Now that we have covered the main parts of AKS, it is important to know about versioning and how it relates to Kubernetes versioning. Let's take a look.

Versioning

An important concept to understand is versioning for AKS. As AKS is based on the original Kubernetes project (and will continue to be so), Microsoft follows the same versioning. This also means that Microsoft cannot indefinitely support all versions of Kubernetes that are deployed to Azure. For this reason, Microsoft uses the following system of versioning and support.

Microsoft Azure only supports the last three Kubernetes minor versions that are GA on AKS – for example, if AKS currently has 1.21 in GA, the versions supported are 1.19, 1.20, and 1.21. That means that you do need to keep up! Changing versions on your cluster requires just a few clicks, but it does need to happen. Running an unsupported version may get you in a whole world of trouble.

As Kubernetes releases about three versions each year, it means you could run out of support within a year's time after deploying your initial cluster. Rest assured that upgrading usually does not impact your environment much, and if there are breaking changes, this is communicated well before the version including the change is released.

In the next section, we will walk through deploying the billing example demo from the previous chapters but on AKS. And, as we are used to by now, we will do this based on a use case.

Your company provides an e-commerce solution to the market. Your solution is built on a number of different Microsoft Azure technologies.

Your product manager informs you that they have purchased licensed code to help them process billing statements. This code is an extension of the existing e-commerce platform and has a number of specific characteristics. The billing statement generation is an API that is being leveraged by the frontend solution and is called every time a user wants to see their current billing and consumption. This API is standalone and pretty small. Your company is currently investing in decoupling the entire application that this API is a part of. In the process, they are containerizing each individual solution and want to run this at scale in Microsoft Azure. The expectation is to run a number of containers, but not all will strictly follow the development standards for containers, as the app modernization process to achieve this might take several years. Technical debt is real. Additionally, we want to be able to control the isolation and communication between containers that might be added in the future. AKS looks like a good solution, as it allows for a lot of customization and works around some missing standards. The concept of all the add-ons and the level of control appeals to you.

It is your job to get this up and running.

Let's get started!

Deploying containers to AKS

Before we start deploying the actual AKS resources, we need that place for our cluster to live again. Let's create a resource group using the following command:

```
az group create `
  --name rg-aks `
  --location westeurope
```

Once the resource group is created, we should see similar output to that shown in *Figure 6.2*, stating that the provisioning of our resource group succeeded.

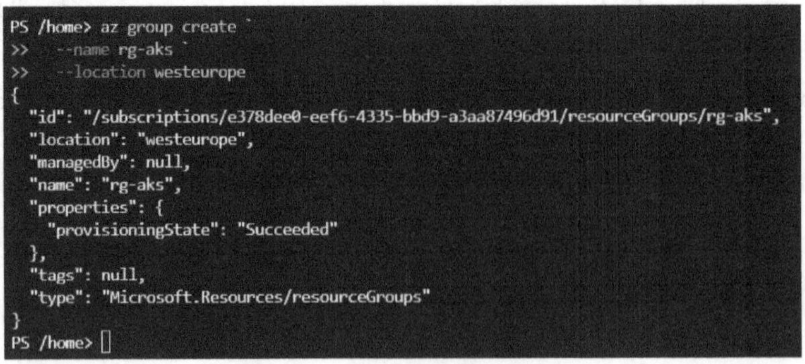
PS /home> az group create `
>> --name rg-aks `
>> --location westeurope
{
"id": "/subscriptions/e378dee0-eef6-4335-bbd9-a3aa87496d91/resourceGroups/rg-aks",
"location": "westeurope",
"managedBy": null,
"name": "rg-aks",
"properties": {
"provisioningState": "Succeeded"
},
"tags": null,
"type": "Microsoft.Resources/resourceGroups"
}
PS /home>

Figure 6.2 – Resource group creation

Now that we have verified that the resource group is there, let's create an AKS cluster! For this, we will be using only a few parameters. However, please note that when you want to add more features and go all out, the set of parameters may grow significantly:

```
az aks create `
--location westeurope `
--resource-group rg-aks `
--name aks `
--generate-ssh-keys
```

This command will create an AKS cluster in the `westeurope` region in your designated resource group. The cluster is currently named `aks`, and we are generating SSH keys that we can use to connect to individual nodes if we have to.

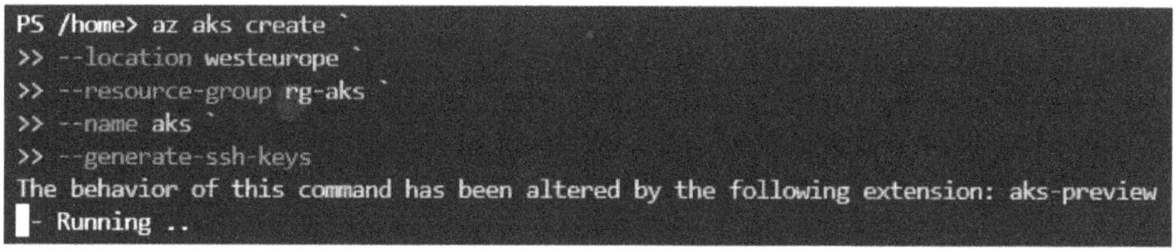
PS /home> az aks create `
>> --location westeurope `
>> --resource-group rg-aks `
>> --name aks `
>> --generate-ssh-keys
The behavior of this command has been altered by the following extension: aks-preview
- Running ..

Figure 6.3 – AKS cluster creation

As you can see in *Figure 6.3*, there is an indicator that informs you whether the Deployment is running. When you perform an AKS Deployment, it will take some time for the environment to be deployed, maybe longer than you are used to with, let's say, app services or storage accounts. As long as the running output remains visible, the cluster is deploying. Depending on the parameter set that you passed, this can take from 2–5 minutes to 15 minutes.

Once deployed, it's time for us to connect to our cluster! For this, we need to retrieve the configuration file that contains our cluster credentials. In traditional Kubernetes, there are multiple ways to grab this file, but Microsoft Azure has made it easy for us. As long as we are logged in and have the right permissions, we can use the Azure CLI to retrieve our credentials:

```
az aks get-credentials --resource-group rg-aks --name aks
```

This results in the context of your cluster being merged into the `.kube/config` directory in your profile. From that point on, you are able to interact with your AKS environment using the `kubectl` command-line tools.

```
PS /home> az aks get-credentials --resource-group rg-aks --name aks
The behavior of this command has been altered by the following extension: aks-preview
Merged "aks" as current context in /home/richard/.kube/config
PS /home> █
```

Figure 6.4 – Retrieving kubeconfig

Now, you might be thinking, "*Merging? I never had a config file to begin with!*" And that is correct! But the configuration file for AKS (or Kubernetes in general) can contain contexts of multiple clusters. Let's say you are deploying multiple clusters – one for production and one for development. These credentials go into the same file, and you can use the `kubectl` command-line tools to switch between contexts/clusters.

> **Important note**
>
> Having multiple *contexts* in a single file might seem like a bit of a security risk, and to be perfectly honest, it can be! This chapter is about the introduction of AKS, and we are not looking to overcomplicate it. Further in this book, we will talk about different levels of security to mitigate this, with solutions such as Azure Active Directory.

Before we can run the container on AKS, we need to tell the cluster what we require of it. We do that in the form of YAML. As we learned from the *Understanding how AKS works* section, the best way to do that is to create a Deployment that contains a Pod. And, yes, the Pod in turn contains the container. The minimum configuration we need to run the Pod can be found in the GitHub repo (`https://github.com/PacktPublishing/Azure-Containers-Explained`) under the `CH06` folder. The file is called `deploy.yaml`.

That looks like a lot! For now, you don't have to remember everything. You can use this file if you want to follow along. We will take you through the important bits.

It all starts with `apiVersion` and `kind`. We're naming these together because they cannot live without each other. Remember when we said that we will use YAML to tell Kubernetes what we require of it? These two properties are key to that concept. By defining `apiVersion`, we are telling Kubernetes which version of the API we want to talk to. Different `apiVersion` instances can contain different features. With `kind`, we are telling Kubernetes what type of object we want to create. By defining these two, we are letting Kubernetes know what to expect from the configuration file (YAML). It's like setting expectations. However, we do need to match those expectations. We cannot simply start adding properties of different kinds; the cluster will return an error if you try to do that.

The next important part is the metadata. We are adding a label with the `name` key and the `billingstatements` value. In the *Understanding how AKS works* section, we discussed the concept of Services and how they load-balance traffic for one or multiple Pods. We are using this label as an identifier for the Service. More on that in a bit!

Looking a bit further into the YAML, we will see a property called `containers` (plural, as a Pod can contain one or multiple containers). This is where we tell AKS where it can pull our container image from and what the runtime configuration is, such as the port number and environment variables.

We then see a separator (`---`). In YAML, this means *end of file – expect a new file*. We can separate the two into distinct files, but for development and testing, you would probably put the Deployment and the Service into a single file. When you start refactoring, however, they might end up in different files.

The next part of the (new) file contains the Service, once again with `apiVersion` and `kind`. Note how these have different values here. We are deploying a Service and we're telling it to be a `LoadBalancer`. This is basically telling AKS to provide an external IP address for our services so that we can access it. And here comes the important part! Remember how we set the label in the Deployment? In the Service, we are now using `selector` with the same value, as we provided the label for the Pod deployment, that value being `billingstatements`. What we have just done is tell the Service, "Hey, we have some Pods running and you can identify them using this label. Please provide the load balancing for any Pod with that specific label."

We can hear you thinking, "I can use this for pushing the next version of my solution and swapping the labels." And you are correct! We will be looking at these scenarios in the next part of this book.

It is time to deploy the beautiful YAML to the AKS cluster! We will now start using the `kubectl` command-line tool that we have mentioned several times.

> **Important note**
> Throughout this book, we will use Azure Cloud Shell, which has `kubectl` preinstalled. If you are executing these commands from a different environment (your local machine, for example), please make sure that `kubectl` is installed. You can do this by executing the following command, `az aks install-cli`, or by visiting the Kubernetes website to download the tools manually: `https://kubernetes.io/docs/tasks/tools/`.

As we said, we have prepared a file for you, which can be found in the Git repo at `https://github.com/PacktPublishing/Azure-Containers-Explained`, so you don't need to copy and paste. The following command will pull the YAML file as discussed and apply it to your cluster:

```
kubectl apply -f https://gist.githubusercontent.
com/PixelRobots/3dab69a2e79e023c68fc0ede38b8e2ce/
raw/5b574b86dc36933bab461898069b50e6d56f7f74/billingstatements.
yaml
```

The command will execute instantly, as all it does is send the configuration to the API and validate it against `apiVersion`.

```
PS /home> kubectl apply -f https://gist.githubusercontent.com/PixelRobots/3dab69a2e79e023c68fc0ede38b8e2ce/raw/5b574b86dc36933bab461898069b50e6d56f7f74/bi
llingstatements.yaml
deployment.apps/billingstatements created
service/billingstatements created
PS /home>
```

Figure 6.5 – Applying the YAML configuration using kubectl

As we used the separator for YAML files (`---`), we are essentially applying two files. The output should be similar to *Figure 6.5*, stating that both files are applied and the resources are created.

Now, we want to verify whether everything we deployed is up and running using the following command:

```
kubectl get deployments,pods,service
```

Using `kubectl`, we are communicating with the AKS API and asking it to return the status of deployments, Pods, and Services that we deployed in the previous steps.

```
PS /home> kubectl get deployments,pods,service
NAME                                READY   UP-TO-DATE   AVAILABLE   AGE
deployment.apps/billingstatements   1/1     1            1           53s

NAME                                     READY   STATUS    RESTARTS   AGE
pod/billingstatements-d446cfd96-nz24z    1/1     Running   0          53s

NAME                        TYPE           CLUSTER-IP    EXTERNAL-IP    PORT(S)        AGE
service/billingstatements   LoadBalancer   10.0.47.36    20.54.218.85   80:30353/TCP   53s
service/kubernetes          ClusterIP      10.0.0.1      <none>         443/TCP        25m
PS /home>
```

Figure 6.6 – Verification of deployed configurations

As we can see in *Figure 6.6*, everything is up and running and our Service is provisioned with an external IP address! It's time to check whether this thing is working:

```
Invoke-RestMethod http://20.54.218.85/BillingStatement
```

By invoking our API, we should have similar results to our billing statements API in the previous chapter.

```
PS /home> Invoke-RestMethod http://20.54.218.85/BillingStatement

dateGenerated          billingStatementId costs currency
-------------          ------------------ ----- --------
4/22/2022 10:08:55 AM                   4  1058 EUR
4/23/2022 10:08:55 AM                  15  2555 EUR
4/24/2022 10:08:55 AM                   9  2460 EUR
4/25/2022 10:08:55 AM                  69  5376 EUR
4/26/2022 10:08:55 AM                   4  3293 EUR

PS /home>
```

Figure 6.7 – Verification of deployed configurations

As we can see in *Figure 6.7*, we can successfully call our API once again. Happy product manager and happy IT ops! For our use case, we determined that the `billingstatements` API will be called from a frontend solution, which is currently being containerized. That means the traffic will remain within the cluster and pushing this to production means we can get rid of the public IP address. For our YAML, that simply means removing the following line and reapplying the configuration using `kubectl`:

```
type: LoadBalancer
```

We can hear you thinking, "*The IP address is not a friendly endpoint to provide to your customers.*" Very true. We need way more features to provide our solution to customers. For one, we want to use a friendly name, such as a hostname. Perhaps we also want to provide a certificate for HTTPS traffic, and maybe we want to inspect some HTTP headers and make choices based on the contents. For that, we need an Ingress resource.

Configuring an Ingress controller and Ingress resources is beyond the scope of this chapter, as it would overcomplicate the concept of *getting a container up and running*. In the next part of this book, we will be looking at production use cases with multiple containers. This is where we will introduce the configuration of Ingress controllers and Ingress resources on AKS.

We don't want to end up with an invoice for unused resources at the end of this month, so let's delete the resources you created during this chapter. We can do this by deleting the resource group. You can do this by using the following command:

```
az group delete --name "rg-containers-aci"
```

In this part, we have introduced AKS as a solution to run your containers on. We have used our use case for the `billingstatements` API to discuss how we can run this on Microsoft Azure's managed Kubernetes offering. Let's now take a look at some pros and cons of using AKS.

The pros and cons of running containers on AKS

As with every technology in this book, there are good things and there are constraints that you need to take into account. Once again, we have documented these as *pros* and *cons*.

Pros

There are quite a few pros with AKS. Let's look at them now.

Managed control plane

One of the biggest pros, if not the biggest, is the managed control plane that comes with AKS. Where it can take a considerable amount of time to deploy a vanilla Kubernetes cluster on your self-managed virtual machines, on Microsoft Azure, all that complexity is abstracted away; you get all of it managed and out of the box.

And that is just for the Deployment. Managing the cluster becomes significantly easier with AKS. We don't have to deal with managing and updating the local etcd store, the certificates, or the agents (kubenet and kubelet). Yes, we still need to press the *buttons* to move to a new version of AKS, but once those buttons are pressed, it's just a matter of sitting back and relaxing. And who doesn't like that?

Easy integration with Azure such as Azure Active Directory

The fact that Microsoft Azure manages the control plane is a huge benefit. But there is more. Running AKS also means you have access to a world of Azure technologies. Microsoft has implemented multiple integrations between AKS and Azure itself. And then we also have community initiatives that also provide similar integrations.

Whether you want to automatically provision DNS records because you deployed a new customer, pull secrets from Azure Key Vault, or protect access to your cluster using Azure AD, AKS has got your back!

Compatibility with the CNCF landscape

When you think *cloud-native*, you think **Cloud Native Computing Foundation** (**CNCF**). A lot of projects are being released and managed through CNCF. Most of them can run on Kubernetes. Think of solutions such as KEDA, etcd (the database powering Kubernetes), and Helm (the package manager for Kubernetes).

In fact, Kubernetes itself is a CNCF project. Having integrations with numerous cloud-native solutions means that the capabilities of Kubernetes are almost infinite! It is impossible to name all of the projects, but if you are interested, take a look at the projects on the CNCF website: `https://www.cncf.io/projects/`.

Upgrades and updates

As mentioned in the chapter multiple times, upgrading and applying updates merely require clicking a few buttons. That is a huge improvement compared to vanilla Kubernetes, where upgrading can be quite a daunting process. Your ops life is definitely easier if you run Kubernetes in Azure!

Cons

Even with all of the aforementioned pros, AKS does have some cons. Let's look at them now.

Steep learning curve

Where the technologies in the previous chapter were pretty much *next-next-finish* deployments and took little effort, AKS is a little different. Even though a lot of the management and complexity is abstracted away because the control plane is Microsoft management, if you want to take your solution on AKS to production, you need to do some studying and practice. The ability to implement customization and different add-ons also means truly understanding the inner workings of Kubernetes, not just to deploy it but definitely to debug everything that is happening! It's not necessarily a *con*, but don't be fooled by its complexity.

A changing node SKU requires a new node pool

When you deploy a node pool, you choose which SKU (VM size) you want to use. This determines the available disk, CPU, and memory you have available per node. There will be a point where you require more compute for a single node instead of simply scaling out to another instance. This will require you to create a new node pool and cordon (drain) and delete the old node pool. If you stick to the concepts of statelessness and scalability that we have discussed multiple times in this book already, this won't be a problem. However, if you have long-running jobs that cannot be stopped and restarted/continued easily, this might be a challenge for you.

Enabling and changing some features requires a cluster redeployment

To be fair, having to redeploy a cluster has been brought to a minimum already, and we expect even more features to allow for configuration during the life cycle of the AKS cluster. However, some settings and features cannot be changed or enabled once the cluster has been provisioned. This means redeploying your cluster if you want to add a feature. The solution to this is infrastructure as code, which you should have anyway, as it is not only great for quickly deploying new environments but also serves as a big part of your disaster recovery scenario.

Summary

In this chapter, we have learned that AKS deployments are pretty straightforward when you are just getting started. However, that's also a big risk. Kubernetes provides an almost endless list of features and requires considerable knowledge and configuration when you want to go to production. But once you get there, the sky is the limit. We see that Kubernetes is definitely an option for workloads with higher demands for resources and capabilities, and you can really configure it to your liking. Microsoft took away the complexity of vanilla Kubernetes by delivering a managed offering on top of it, making Kubernetes accessible to a broader audience. The integration capabilities that Azure provides with AKS are what makes it stand out compared to other public clouds with a managed Kubernetes offering.

In *Part 2* of this book, we will be diving into actual production use cases, ranging from businesses in a start-up phase to businesses that are about to scale from the **Small-to-Medium Business** (**SMB**) to the enterprise segment. For each use case, we will dive into the technologies and their more advanced features to support these types of businesses. Stay tuned!

Part 2: Choosing and Applying the Right Technology

Now that you have a basic understanding of the different technologies, how will you put them into practice? What does this mean for your architecture and the future development of your solution? In Part 2 of this book, we will discuss the short-term and long-term capabilities, possibilities, and decisions that will impact your options for the future.

This part comprises the following chapters:

7
The Single Container Use Case

In the first part of this book, we covered the different technologies you can use to run containers on Microsoft Azure. There, we provided a brief introduction to those technologies. Now, it is time to look at the different use cases when you have just one container to run.

As we have pointed out multiple times, whether a technology is the right fit for you depends on your use case. If you match the right technology with the right use case, success awaits you. In this chapter, we will dive into different use cases and see which technology would be the best fit from a technical and business perspective. Yes, also from a business perspective. In the end, everything you do as an IT professional aims to add value to your company and its customers. If a technology does not provide any added business value, why even bother?

In this chapter, we will cover the following topics:

- Scenarios and types of solutions
- Selecting the right Microsoft Azure technology
- Decisions and the future of your solution

Scenarios and types of solutions

There are many reasons for just using a single container. But really, what does it mean when we say, "Running a single container?" This needs some clarification. In the previous chapters of this book, we have seen and explained the basics of containers but also the basics of *Pods*. Remember Pods? These are the things we deploy to solutions such as Azure Container Apps and Azure Kubernetes Service, and they can contain one or multiple containers. We also discussed that you can run multiple containers using solutions such as *Docker Compose* as a single deployment.

In these scenarios, we are not necessarily sticking to *just one container* but to one, perhaps two, containers in a single deployment, which means they are tightly coupled. We will discuss why that is shortly when we discuss the scenarios we come across most often.

> **Important note**
>
> We have provided lots of technical examples in the first six chapters of this book. As we are talking about single container use cases here, we will not be repeating the technical content as we used the single container example in the previous chapters, which covered the how-tos. From a technical perspective, we will pick up where we left off in *Chapter 8*.

Building a new solution – starting small

From an IT professional or developer perspective, the best thing that can happen to you is a greenfield scenario where you can start selecting technologies and writing code without any existing constraints. Of course, you still need to follow technical and financial requirements, but there is no existing infrastructure holding you back. Most of these solutions start very small, either in a single web app, a single Azure function, or, in this case, a single container that contains the business logic to do what you require from it.

You could start writing code and run it locally, but at some point, the solution needs to run on infrastructure in the cloud. This means you need to think about what technologies fit your use case, as well as what the future is going to bring. Will this remain a small solution for the next year or are you planning to add functionality (and thus, containers) within the next few months? This is an important consideration because while all the technologies we've discussed can run a container, not all technologies can scale to a size you might require from it.

You're migrating to a container-based infrastructure

Another very common scenario is where you have an existing solution, and you require this solution to run in a container. This is where things get a little tricky. You have something that already exists and works, and you need to containerize and run it. This is commonly called brownfield.

We've already discussed Windows versus Linux containers, and the runtime you are currently using is very important when making that decision. For example, many (legacy) solutions still run on **.NET Framework**. That will not fly well in the world of Linux containers as .NET Framework is simply not cross-platform. This means you need to choose whether you want to refactor the code and use a more modern runtime such as .NET 6 or run a Windows container.

How do you make that choice? Well, you could say that the easiest way to go about this is to containerize the solution using Windows. However, that will already limit you in terms of technology/platform choice. Not all platforms support Windows containers and for the platforms that do, considerably more overhead is allocated for running Windows containers as opposed to Linux containers. This means that in the end, running Windows containers is more expensive not only financially but also operationally than running Linux containers.

What does that mean? It means everything! There's not just a technical choice to be made here – there's also a business choice. Let's look at a scenario.

> You have an API that currently runs on Azure App Service and is built on top of .NET Framework. The solution started small and is deployed as a single-tenant solution for each customer. The solution has proven to be very successful in the market and there are plans to expand its functionality and introduce additional APIs, a frontend, and a more scalable backend.
>
> Your company has decided to embrace a container-first approach as they are looking to be more cloud-native and leverage flexibility, scalability, and the cloud-native ecosystem. The leadership strongly believes that moving to a container-based platform will greatly benefit the business and they are willing to make considerable investments.
>
> It is your job to get this done.

This is a very common scenario a lot of start-ups deal with. There is a growth plan, which requires technical innovation. Let's take a deeper look at this example:

- The current solution is a single-tenant API that is deployed per customer.
- The solution runs on .NET Framework.
- The company wants to expand the existing solution and add features.
- Cloud-native and containers are the way to go.

Now, as a technical professional, you might be wondering why you should care about all the business-related stuff. You just need to containerize this thing, right?

Wrong. You need to understand the business requirements before you can make the right decision. As mentioned previous, you have two options:

- Refactor the code to Linux containers and have more choice in terms of technologies and features.
- Containerize to a Windows container and be done with it.

Which one you choose greatly depends on several factors: how well you can bring the story across to your leadership team and whether there is a long-term focus (is the company willing to invest now to be better off later?).

Containerizing to a Windows container takes significantly less time (and thus, money) to do as opposed to refactoring the code to use .NET 6 and running a Linux container. However, the costs/investment of that refactoring process may very well be worth it in the long run. The main question is, how much technical debt are you willing to accept and what would it cost you to fix it in the future? This is the business case that requires your input and will lead to your technology choice.

You are extending your current solution to use a container

Chances are you already have existing infrastructure and instead of re-platforming, as we discussed in the previous paragraph, you just want to add a new feature by introducing a containerized solution.

As you learned in the previous chapters, if you are going with a single container, pretty much any technology in Azure goes. However, let's picture a scenario.

> You have an existing solution that runs on Azure App Service, and you provide your customers with two APIs hosted in two separate app services. Another division within your company has developed a container that processes CRM data from the customer once a day and stores it in a database that is accessible by the two APIs. Everything is deployed as a single-tenant solution, which means each customer will have two app services, one database, and soon, one container.

From a technological perspective, you have a couple of choices. The likely platforms would include the following:

- Azure App Service
- Azure Container Instances
- Azure Container Apps
- Azure Functions
- Azure Kubernetes Service

You can rule out Azure Kubernetes Service pretty much straightaway. From a financial perspective, this would not be very interesting and from a technical perspective, overkill. Azure Container Apps ideally requires development effort to leverage the built-in technologies such as Dapr and KEDA, which is not always ideal.

Then, you are left with Azure Functions for containers, Azure App Service, and **Azure Container Instances** (**ACI**). In the example, you saw that the CRM data is only processed once a day. Since Azure App Service and Azure Functions for containers both require an App Service plan that is billed for a fixed price per month, whether you use it or not, this wouldn't fit the use case from a financial perspective. ACI, however, only runs when we need it to. You're only paying for that piece of processing once every 24 hours.

From a financial perspective, you can go with ACI. As ACI also comes with network integration, you can even manage the outgoing traffic to the customer CRM through a NAT gateway if they require a fixed IP address in their firewall configuration.

Do keep in mind that this is a perfect solution if you are just adding one or two containers to your existing solution. If you come to a scenario where the majority of the solution will be running in containers, it's time for the next phase; think about migrating the app services to containers and look for a better technical fit.

In this section, we briefly looked at the common scenarios for solutions that often run in a single container:

- Building a new solution from scratch
- Migrating/containerizing an existing solution
- Extending your existing solution

Let's dig a little deeper into those scenarios and go through the decision-making process of selecting the right technology.

Selecting the right Microsoft Azure technology

So far, we have seen that there is more than just a technical decision to be made and that there is always a business and financial component to it. Let's take a look at a new use case.

> You work for a company that provides an e-commerce solution. Customers come to you when they need anything in the form of a web shop. Among other features, the e-commerce solution allows the end user to order items, perform payments, view their order history, and review items they have purchased previously. The solution is single-tenant, which means a separate instance or a combination of resources is deployed for each customer. Technically, the solution is monolithic and runs on **Internet Information Services** (**IIS**) on **virtual machines** (**VMs**). The current runtime is .NET Framework 4.8. Migrating to a more modern runtime is on the roadmap but is currently not a top priority.
>
> Recently, customers have been asking for a connection between the e-commerce solution and their CRM system. Product management has decided to allocate development resources to build an API that can process this information and ensure it can be called by the CRM system whenever the customer requires it.
>
> It is up to you to recommend the correct technologies, help the developers get started, and deploy the solution.

So, we have this use case, and you could say, "Just build another .NET Framework solution and run it on the IIS environment." Well… let's not extend the current solution by adding more technical debt. First, let's break down this use case. What do we know that can help us in our decision-making?

- The solution is single-tenant.
- The runtime of the e-commerce solution is .NET Framework 4.8.
- Modernizing is on the roadmap but not a priority.
- A new feature is to be added that requires development.
- The API needs to be called by a CRM system.

Those are some interesting facts, but what can we learn from them? Surely this isn't all the information we are going by? And that's correct. As an IT professional, you also have requirements for any new solution to be built. The good thing here is that you are involved before the API is developed; you can now influence the development process – something that is very important when you are talking about containers where infrastructure and development impact each other. Let's see what else we need to think about:

- How is the solution going to scale when we receive a lot of requests?
- What runtime are we going to use?
- Is the solution going to hold any state?
- What about security?
- What technology would be the best way to go?

Not everything in this list is something to be answered by a single person, but they are important questions in the decision-making process for this use case. Let's go through the things we know and want to know step by step and see whether we can make a decision.

Security

Always think about security when building a new solution. Of course, there are different layers of security, and we are not getting into the development part of this. But what we do know is that the API is going to be called from the customer CRM. That should raise a question related to security: who is going to request what information and where is it going to end up?

We know that the e-commerce solution is single-tenant, so we have the choice to write a multi-tenant API. Even though it is very popular to develop multi-tenant solutions and it can bring down overhead and improve efficiency, it would not be the best choice in this scenario. The complete e-commerce solution is single-tenant and written in an older framework. Adding a multi-tenant API that can be accessed by multiple customers means we need to differentiate and make sure the customer can only access their backend when querying the API. It is possible but it will overcomplicate things. Ideally, this will also be a single-tenant solution that is deployed along with the other resources for the customer to maintain the isolation that is currently in place.

Runtime

We know that the current runtime is .NET Framework 4.8, and we don't want to introduce any technical debt when writing new code. It would be nice to be cross-platform since we don't know what we will do in the future. The best choice here is to go with a runtime such as .NET 6. It is cross-platform, which doesn't limit the technologies that we want to use, and developers are already familiar with .NET Framework.

State

You *always* want to know whether the solution is going to hold any state. If there is a persistent state, it means that after restarting the solution, the state needs to be there. It is also likely that you need to introduce a backup policy to make sure that the data can be retrieved. Luckily, when a solution is still to be developed, we can influence this.

What we know is that the current e-commerce solution has a database (per customer, as it is single-tenant) that sounds like a state. We can assume that our new API can access the database. And, we know that should be able to request pretty much any data through the API – data that has to exist in the database. Therefore, we can conclude that our new solution will not hold any state.

Usage and scalability

The thing with APIs is that we don't know when the customer is going to call them. It is very unlikely that this would happen every single second. However, this behavior will be unpredictable. Some customers may only request small sets of data every hour, while others might request a lot of data every night at 11 P.M. to update the records in the CRM system. Either way, it is unpredictable.

Decisions, decisions, decisions

That was a lot of information; now, it's time to make decisions. Essentially, what we did in the previous paragraphs was abstract away the requirements. Let's see what they are:

- The solution has to be written in .NET 6.
- The solution needs to scale on demand as its usage can be unpredictable.
- The solution needs to be single-tenant.
- The solution needs to be accessible to the customer.
- The solution needs to be portable as modernizing the solution is on the roadmap but currently isn't a top priority. We don't want to write something that is going to impact future choices too much.

From a technology perspective, we know we don't want to deploy an additional solution on the existing platform that is currently hosted on IIS with .NET Framework 4.8. Yes, technically we could, but that means we would be introducing a new solution with an old framework and that we also have that requirement of scalability. If the API is going to use a significant amount of resources, we don't want to scale the entire VM on demand. This would impact user experience (VM restarts) or it would require us to scale up the machine and have unused resources we are paying for sit there until the API is called.

We also don't want to commit to any specific technology and lock ourselves in, essentially limiting the future choices for the platform. From a business perspective, we want this to be as scalable as possible, as cheap as possible, and, again, as portable as possible.

This rules out Azure technologies such as Azure App Service, where we have to allocate an App Service plan and pay for the fixed amount per month. It also rules out Azure Functions, where we are then required to write specifically for that platform, which means code-specific for Azure Functions. That's not very portable either.

Containers, on the other hand, are the perfect technology! We can develop the solution in .NET 6 and thus use a Linux container as the runtime is cross-platform. Additionally, we can just write the code as we want, containerize it, and, in the future, use the same code base and deploy it on another solution (if, for some reason, we want to stop using containers).

So, containers it is! Now, we just need to pick the right platform.

As we have seen in the previous chapters, we have multiple options when it comes to running containers in Microsoft Azure. Will we go with containers in Azure Functions, ACI, or Azure App Service? Let's take a look.

Azure Functions for containers

That is a big no. Containers for Azure Functions don't fit the use case as we would have to develop specifically for Azure Functions as this is also required when using a container. Additionally, we need a premium plan, which means a fixed price per month. This doesn't make it very scalable from a financial perspective.

Azure App Service for containers

Again, this is a no from our side. As with Azure Functions for containers, it requires an App Service plan and doesn't fit the use case.

Azure Container Apps

Now, we wouldn't say *a big no*, but it's still a *no*. This comes down to future decisions about the platform and our development capacity. Yes, we can simply host a container on Azure Container Apps but to effectively use it and leverage its power, we need to introduce technologies such as Dapr and KEDA to our developers. For one, this would require additional time and a steeper learning curve for our developers. Additionally, we don't know what the future roadmap of the platform looks like. If we are talking about multiple containers, this might be an option, but not for now.

Azure Kubernetes Service

This one speaks for itself. It's a marvelous product but somewhat overkill from a technology perspective. We could still argue that from a financial perspective it might not be that bad of an idea as the solution is single-tenant and 100 customers means 100 containers, and that argument is very valid. However, we are still talking about adding a single feature (the API) to an existing solution that runs on IIS and a VM. Adding the complexity of Azure Kubernetes Service to the day-to-day tasks of the IT professionals in our team is something we need to think about, especially if it differs from the technologies they are currently used to managing.

Azure Container Instances

Now, this is interesting. We require on-demand scalability, which is essentially what ACI provides. It is a serverless container runtime and we are only paying (per second) for it when it is executed. ACI also provides extensive networking features, which means we can completely silo the single-tenant deployment.

An extra argument for using ACI is its simplicity. Where platforms such as Azure Container Apps and, in particular, Azure Kubernetes Service require more knowledge of the *inner workings* of the technology, with ACI, we can pretty much just have the developers write the solution and provide them with a Dockerfile, and our IT professionals can use their existing Azure and Ops knowledge to get it up and running without a lot of studying. We saw this in *Chapter 4*, where we deployed a container using just a few Azure CLI commands.

Based on this use case, we can conclude that ACI is the best option but the technology itself doesn't matter. It's the process of thought and having the right mindset to achieve the result: a decision on how to deploy our single container solution. It's not even about the *container* in particular – we have determined that containers are the right way to do this, and we have motivated every single step based on the requirements. But we have now made a decision: one with minimal impact on future decisions (hopefully). Whether that is true, we will see in the next section.

Decisions and the future of your solution

Will the decisions we made in the previous use case come back and haunt us? Hopefully not. But what's important is that technologies evolve fast. Roadmaps are not set in stone and things change. This is how we concluded that we should use ACI for our use case.

We could assume containers are going to be very successful and we probably even have some insights into the company's future that others don't. But simply going for the platform that provides all the container capabilities we would ever need (Azure Kubernetes Service, for example) might cost us a lot of effort in terms of training developers, expensive infrastructure, and one or two sleepless nights to get it done.

Small steps will eventually become huge leaps. We have just made a small step by introducing container technology to our developers but with minimum requirements. They can still write code as they are used to, and we will help them create that Dockerfile, which isn't that complex.

Essentially, what we are doing is not only making the right financial and technical choice – we are gradually introducing container technology into the organization and providing a learning-on-the-job experience.

But what if the scenario repeats and another API is introduced? Did we make the right call? What if a year later, new technical decisions have been made or that roadmap finally becomes clear? Did we make the right decision? Maybe we did, maybe we didn't. Let's see what *could* happen.

There are plans to develop another API

Product managers don't sit still. Another request came in to add functionality to the existing solution and the decision was made to not develop any new features on the current .NET Framework 4.8 version of the solution. Luckily, you already went through a big part of the thought process, essentially shaping a part of that future roadmap. You made the right call, and everyone is now waiting for you to make the next call. Will it be a container that runs on ACI or not?

Your first and foremost question should be, "how likely is it that a third solution will be added in the future?" Hopefully, your peers can answer that question. The path of least resistance will be to add another container to your current ACI, not requiring any additional knowledge or learning on the job from your colleagues. But if the answer to the question is "chances are big," then maybe it is already time to start thinking of technologies such as Azure Container Apps.

Again, that would require some learning on the job, but not as much as when you haven't been running any containers yet. People are already somewhat used to it. What you would be doing is providing them with more features. It is like saying, "Hey, we have been using containers for the CRM integration, but if we are going to another one, perhaps we can explore this other technology and provide you all with more features."

This is where your choice of using ACI is once again validated to be the right one. Deploying to ACI is very straightforward and would have not taken more than 1 or 2 days of your time to get it right. Along the way, everyone has gotten a better understanding of containers – that is the basic knowledge you will always need, regardless of the technology. You could say that every moment spent with ACI was the right investment for the future. So, you have two options:

- Move to Azure Container Apps and invest some more time.
- Stick with ACI, which would require no additional investments other than writing code.

All in all, this confirms you did the right thing!

The roadmap dictates Azure App Service is the way to go

You are a year from now; you still have your VMs, IIS, and single-tenant containers running. Leadership is convinced containers are not the way to go – they are certain everything needs to run on Azure App Service. That could be from technical reasoning (.NET Framework support) or because they were convinced otherwise. Either way, sometimes, a decision is out of your hands.

However, you did make the right call using containers. If we go back to our use case from the *Selecting the right Microsoft Azure technology* section, we decided to use .NET 6 as the runtime and containerize the API. Containerization only took a Dockerfile and you invested 1 or 2 days to get it up and running.

The only investment you would be throwing away here is 2 days tops. In a year, that isn't too bad. But imagine you had spent weeks getting the perfect Azure Kubernetes Service environment up and running. That would be quite the time thrown away!

Once again, the right call.

The roadmap dictates everything will run in containers

In this case, you would be jumping for joy! Your decision had an impact that convinced leadership that containers are the way to go. This would mean a lot of rewriting for your developer colleagues, but your minimal effort to extend the current solution with a container paid itself back very quickly. It was time well invested. Once again, you have some decisions to make. Will you go for the all-out configuration and use Azure Kubernetes Service or are you still taking small steps?

Once again, this completely depends on the roadmap. This is where you have to start thinking about development capacity: Ops capacity and training.

Whatever you choose as your next step, the business component is very important to understand. As an IT professional and enthusiast, you probably want to get started tomorrow and make that big leap to an all-out containerized solution and allocate all possible resources for this. However, you still have customers. Allocating all those resources results in less time spent keeping the lights on in the current platform. Modernizing a solution like this takes time and has a huge impact on software architecture. You cannot afford to take away too many resources from the currently running business. It is in no way acceptable to tell a customer, "Please wait a couple of months; we will add that feature by the end of the year."

What you need to understand and never forget is that redesigning and refactoring a big solution will add value to the customer, but not while you're doing it. It's the result that benefits the customer. The process of refactoring or rebuilding does not benefit the customer (in their opinion). It might even degrade customer happiness if you don't allocate your resources carefully.

You might be thinking, "why are you telling me this?" Well, if customer happiness is degrading, it will impact everyone and put unnecessary pressure on people. After all, it's the customers who make sure you have a business at all.

So, once again, you have a choice to make: go all-out and do everything at once or keep taking the small steps. From a technical perspective, you have two options here:

- Refactor and release everything at once, and then migrate customers over to the new platform.
- Decompose the solution into smaller pieces and refactor piece by piece.

Choosing one over the other might have a different technological impact. Going for the first option means that Azure Kubernetes Service is a very good option but comes with the risks described previously.

Option two usually wins the popular vote. This also means you can introduce new technologies along the way. Running the first couple of containers on Azure Container Apps is a very viable solution. Whatever is rewritten to .NET 6 can be containerized and can leverage the Azure Container Apps technologies such as Dapr and KEDA. Once you have a couple of them running and have gained momentum, you can move to Azure Kubernetes Service. This will result in your infrastructure gradually advancing at the same speed as your developers refactor the solution.

However, there is always the question, "But what does that look like and how do I do that?" In the upcoming chapters, we will go through the scenarios for the different container technologies on Azure based on option two: decompose the solution into smaller pieces and refactor piece by piece, and go from a small container infrastructure to an enterprise-level infrastructure based on practical examples.

Summary

In this chapter, we took you through the decision-making process of going from no containers at all to a single-container solution. You saw that there is no simple answer to "Do I use technology *X* or *Y*?" Instead, you have learned that even deploying a single container will have a significant impact on your options for the future, whether that is from a business or a technological perspective.

No matter how simple the question is, it is important to dissect it, ask questions, and find the right technology now and in the future. We explained how selecting a specific technology now doesn't mean you have to stick to that technology until the end of time.

In the upcoming chapters, we will show you the technical execution of those choices.

8

Deciding the Best Fitting Azure Technologies for Multiple Containers

In the previous chapter, we went into detail on how to run a single container on Azure. There are a lot of use cases that fit those technologies but what if we are planning to use more containers?

Not all technologies on Microsoft Azure are best suited to support those workloads, and each workload could have differences in requirements, which will greatly impact the choices we have. Sometimes, it may even feel like it doesn't matter which technology you select, but the reality can be different. In reality, it is usually more than the superficial features you see or the ballpark figure in costs you made. Sometimes, you just need to experience what is best for your solution. We will try and help you with those choices by going through a few examples and mapping them to some use cases.

In this chapter, we're going to go over the following topics:

- Multiple container scenarios and their solutions
- Deploying the right Microsoft Azure technologies

Multiple container scenarios and their solutions

The scenarios in the previous chapter still hold true. You could still be starting with writing from scratch but have already planned to deploy multiple containers, or maybe you're rearchitecting parts of an existing solution to run in containers. No matter which of the two scenarios you have chosen, there are some important technical aspects we need to take into account:

- How *dependent* are those multiple containers on other containers during runtime?
- Will those containers interact with each other, and if so, how?
- Are we concerned about latency between those containers?

These are some very good questions that are often forgotten. They're not forgotten because they're not important; they are forgotten because it doesn't fit the mindset of traditional infrastructure and architecture. Let us explain why.

Having a greenfield scenario where you are writing a new solution from scratch sounds great but is often not the reality. In fact, we are usually confronted with existing code that has been maintained for many years. A commonly occurring scenario is where a small group (or sometimes even one person) started writing software, brought it to the market, and started adding features once the product became more popular. There usually wasn't any form of architectural design being created while writing the first version. When bringing a product to the market, the code quality and architecture aren't usually on top of the list; it is purely functionality driven. That is fine, this is how most software solutions start. But that also means you are being confronted with some technical debt.

We often see that, in these scenarios, all components within the software are heavily dependent on each other; you change something in the frontend and something in the backend goes wrong. If Services and code are that tightly coupled, imagine what changing the infrastructure (containerization) will cause. Where, traditionally, everything runs on the same virtual machine, the same hardware, and within the same instance of an operating system, on a public cloud, that can be quite different. Even on virtual machines, you are normally not talking to a physical disk that is attached to the machine you are running your solution from. In fact, you have to take the network traffic between the operating system and the disk into account. What if there is latency? What if the disk is temporarily not available? You have to take these scenarios into account. This is because even though the public cloud makes a big promise, it does require you to rethink the way your solution interacts with infrastructure.

What we need to understand is that in a public cloud and with containerization, in particular, infrastructure can kick start your innovation or can be severely limiting if you choose the wrong components.

Let's take a look at some technical scenarios we often come across when it comes to running multiple containers.

Migration of the tightly coupled solution

This is probably the scenario we come across most often. Even though the common answer to the question *Can it run on containers?* is usually *yes*, in reality, it usually requires a bit more work. Let's say you have a solution that consists of business logic and, additionally, you have a self-written logging engine that is accessible to the solution over `http://localhost`. The solution running the business logic and the logging engine run on the same virtual machine. You want to containerize this solution. What we have learned in the previous chapters is that containers (or Pods in the world of Azure Kubernetes Service) can run on different machines (nodes). That behavior is problematic for a solution that requires a logging engine to be available on the local machine. What we often see is that there is a technical debt when it comes to synchronous versus asynchronous operations in code. It is not unlikely that you will come across a solution where interaction with the logging engine is synchronous. What this means is that the code is waiting for the logging operation to finish before continuing. If we were to containerize the app and the logging engine into two separate containers, one for the business logic and one for the logging engine, and they were to be deployed as individual objects, we might introduce latency, which causes performance issues for the solution. So, why do we elaborate on this scenario? The answer lies in the capabilities of the technologies we can use and how we can use them. A clear requirement from this scenario is that both containers would need to run as close together as possible. In the world of Kubernetes, we would create one Pod with both containers in it. In Azure Container Instances, we use a container group containing multiple containers, and in Azure Container Apps, we can define multiple containers in a single app. That also pretty much describes the technologies that are the best fit for these scenarios.

However, there is a downside to tightly coupling. When containers are tightly coupled and thus have to be deployed in a group such as we just discussed, it also means they share some characteristics of the technology being used. Now, it's not to say that these downsides are necessarily a bad thing or should prevent you from doing this at all but it's important to know what they are. They have been listed as follows:

- They share **scaling rules**, which means if one container needs more resources, both containers will get more resources.
- They usually restart together when you update the Deployment.

What we can see is that the main downside in this scenario is *if you touch one container, you touch the other one*. And, as long as you keep taking that into account when you are scaling, restarting, or updating, then tightly coupled containers are perfectly fine. It is important to note that even though migrating a tightly coupled solution to container technologies will bring benefits in terms of speed and repetitive Deployments, it doesn't solve the so-called *monolith problem*.

Quite often, we see a migration like this as the first step to containerization. However, we always recommend that you don't stop here. Your next step should always be to decouple the solution to benefit even more from container technologies.

Decoupling a solution

Decoupling a solution does not simply mean *let's pull this thing apart and deploy it in two different Pods*. There is a lot more to it and it may even require more technologies. It's also not really a *container* question here and the actual tasks to decouple your monolith are a bit out of scope for this book, but we need to address it still. As having tightly coupled solutions reduces the capabilities of our container technologies, as we have discussed in the previous chapter, decoupling them will open up a world of capabilities.

Basically what we are saying is to *prevent two containers from being directly connected or dependent on each other*. What this means is that if we want to restart one container, or even take it offline, the other one should remain operational.

For example, let's consider an e-commerce solution with a webshop as a frontend, and a second solution that provides an API for the order processing backend. We want to be able to update, turn off, or restart the order processing backend without causing issues in the frontend. If we can achieve that, that means that we can independently scale the frontend and backend, only allocating resources (and thus spending money) where we have to. Additionally, because we can restart them individually, we can also update each container individually, allowing features to be added without impacting the entire solution on update.

Now, decoupling a solution as described previously does not mean they operate independently from each other. From an infrastructure perspective, they are, but from a code perspective, they might still be tightly coupled. From a development perspective, if the frontend solution sends synchronous operations to the API, this still means the frontend would *suffer* if the backend is not available. To overcome this, we would ideally introduce a messaging solution such as Azure Service Bus to store messages that are used for interaction between both. Containers are a perfect solution to introduce such Service; we can even use solutions such as **Kubernetes Event-Driven Autoscaling (KEDA)** (which comes with Azure Container Apps by default and can be added as an add-on with Azure Kubernetes Service) and we can build the optimal integration between code and infrastructure. We will take a look at this in *Chapter 9*, where we will create an infrastructure design and implement it using the right technologies. In short, we can say that decoupling truly takes both infrastructure and development efforts.

The Windows and Linux mix

This scenario is sometimes frowned upon; it is sometimes complicated, and sometimes it can be the amazing start to your containerization journey.

Greenfield scenarios are great; we can start from scratch and build whatever we want. However, the reality is quite different. It's not often that you get to work with a startup where you can design anything and implement anything you want. In fact, usually, we are confronted with a customer that has a pre-existing application and wants to move to the cloud. A lot of the time, this solution runs on Windows.

With the growing support of Windows container support on Microsoft Azure, running Windows containers becomes more common. Across the customers we have worked with, we observed an increase in the use of Windows containers – not as a permanent solution but as a means to an end. Let us see why by using a use case.

> A software company has been developing a solution for the past 20 years. This solution runs on Windows IIS and heavily depends on the operating system itself. Over the years, they have introduced new services and new software runtimes such as .NET 6, but a majority of the solution still runs on .Net Framework. They have decided they need to migrate from traditional virtual machines to containerized technology – not because it sounds like a great idea to run everything in Windows containers, but because parts of their solution already run on .Net 6 and can run in a Linux container just fine. Their end goal is to run everything in Linux containers but they are not quite there yet and need to invest significant development resources into taking away the dependencies provided by Windows.

There are two paths this company can take if they want to run their solution on containers:

- Rebuild the current .Net Framework solution to run in .Net 6 and spend significant time and delay the project by months or even a year.
- Containerize their Windows solution as well.

The first option would mean the project is on hold and they will not experience any benefits of containerization or the cloud at all. The second option, however, gives them the opportunity to migrate and gradually modernize their .Net Framework solution.

It is very important to understand the constraints of Windows containers. Even though more Azure technologies are supporting this, they do behave quite differently from Linux containers, as we have seen in earlier chapters in this book.

We see the usage of Windows containers as a *migration* or hybrid scenario. While migration might take up to a year or even longer, it is a much better option than rebuilding everything first, as these projects usually take much longer. This would also ensure that the projects are not being pushed because of other priorities and end up not even being finished at all.

Now that we have gone through the scenarios and looked at a bit of background into each one, let's look at what it takes to deploy them both.

Deploying the right Microsoft Azure technologies

So, how does each scenario work technically? How can we deploy tightly coupled and loosely coupled containers, and can we technically spot the difference? In this chapter, we will deploy both types of solutions. Don't worry, we will revisit the Linux and Windows container hybrid in *Chapter 15, Azure Kubernetes Service – The Next Steps*. For now, let's use the following scenario and see how we can answer the preceding questions and determine which Azure technology is the right fit.

A software company provides an e-commerce shop that allows its customers to buy licenses. For the billing part, they have built an API that is leveraged by the frontend e-commerce shop itself or, in some cases, is directly available for end customers to integrate with existing systems. They have recently containerized their solution and now want to run it in Microsoft Azure. They are still unsure whether they should deploy these containers tightly coupled together or completely independent from each other. Luckily, modern container technologies allow them to test this very quickly.

To demonstrate and test this, we are using the following two technologies:

- Azure Container Apps
- Azure Kubernetes Service

Let's see whether we can get this up and running, test this, and see the technical differences that we described in the previous paragraphs.

Tightly coupled

If this is your first time looking at **Azure Container Apps** (**ACA**), we would recommend you go back and read *Chapter 5*. It will give you a detailed understanding of what ACA is and how it works and will explain the commands that you will see here in more detail.

First, we need a resource group for ACA resources to live. You can use the following command to do this:

```
az group create `
  --name rg-containerapps `
  --location westeurope
```

You should instantly see some output similar to that shown in *Figure 8.1*:

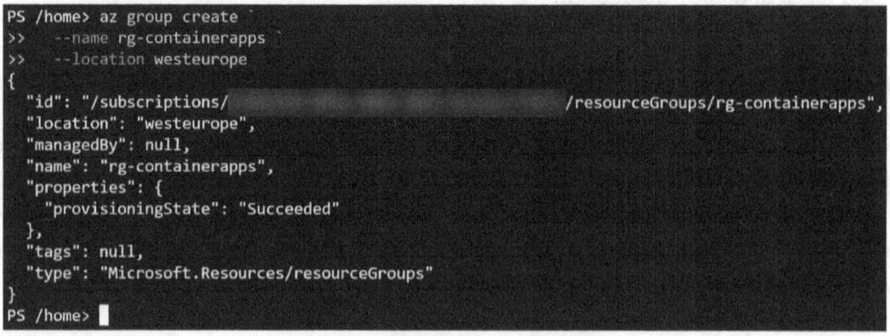
PS /home> az group create `
>> --name rg-containerapps `
>> --location westeurope
{
"id": "/subscriptions/[redacted]/resourceGroups/rg-containerapps",
"location": "westeurope",
"managedBy": null,
"name": "rg-containerapps",
"properties": {
"provisioningState": "Succeeded"
},
"tags": null,
"type": "Microsoft.Resources/resourceGroups"
}
PS /home>

Figure 8.1 – Resource group created

With the resource group created, we can now deploy an ACA environment for this scenario. You can use the following command to create it:

```
az containerapp env create `
  --name tightlycoupled `
  --resource-group rg-containerapps `
  --location westeurope
```

If this is your first time creating an ACA environment, you may be asked to install the `containerapps` extension. It is alright to do this, you just need to type *y* and press *Enter*. After a short while, you should have some output similar to *Figure 8.2*:

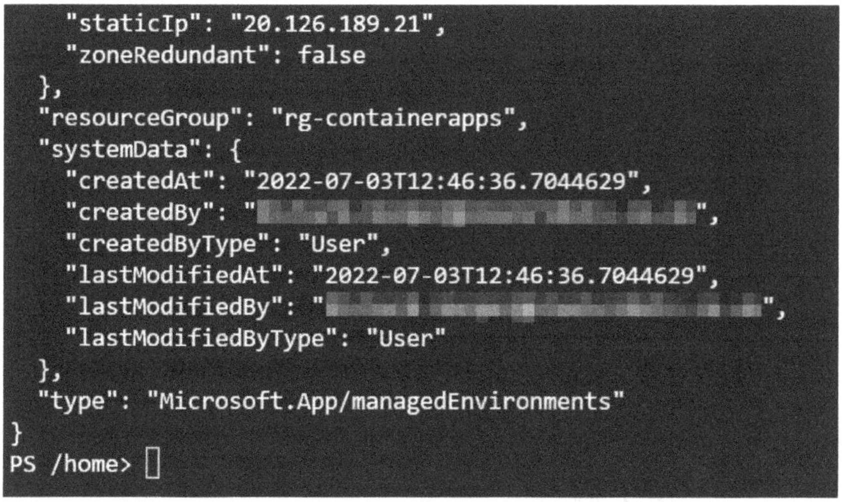
"staticIp": "20.126.189.21",
"zoneRedundant": false
},
"resourceGroup": "rg-containerapps",
"systemData": {
"createdAt": "2022-07-03T12:46:36.7044629",
"createdBy": "",
"createdByType": "User",
"lastModifiedAt": "2022-07-03T12:46:36.7044629",
"lastModifiedBy": "",
"lastModifiedByType": "User"
},
"type": "Microsoft.App/managedEnvironments"
}
PS /home>

Figure 8.2 – Container app environment deployed

Now that we have the resource group and the container app environment, we are ready to deploy two tightly coupled containers. For this, we are going to need to use a YAML file, which can be found in the GitHub repo (`https://github.com/PacktPublishing/Azure-Containers-Explained`) under the `CH08` folder. The file is called `tightlycoupled.yaml`. Create a copy of the file locally as it is needed in the next step.

If you have looked at a Kubernetes manifest file before, you may see some similarities, but they really are different. For you to use this file in your subscription, you will need to add your Azure subscription ID in the `managedEnvironments` value. Once you have the file ready, it is then time to create the container app.

We will call our container app `tightlycoupledcontainers`. You can use the following command to do this:

```
az containerapp create `
  --name tightlycoupledcontainers `
  --resource-group rg-containerapps `
  --environment tightlycoupled `
  --yaml ~/tightlycoupled.yaml
```

After some time, the container app will be created and you will have some output similar to that in the following screenshot:

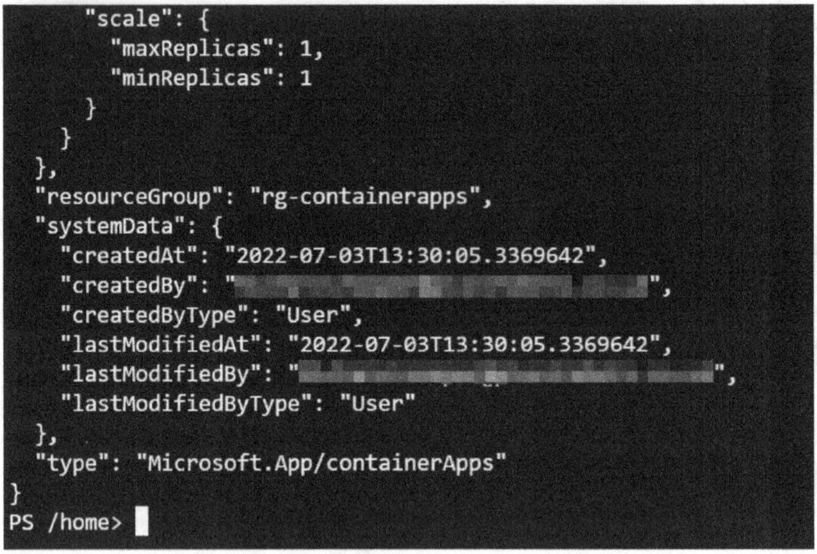
"scale": {
"maxReplicas": 1,
"minReplicas": 1
}
}
},
"resourceGroup": "rg-containerapps",
"systemData": {
"createdAt": "2022-07-03T13:30:05.3369642",
"createdBy": "[redacted]",
"createdByType": "User",
"lastModifiedAt": "2022-07-03T13:30:05.3369642",
"lastModifiedBy": "[redacted]",
"lastModifiedByType": "User"
},
"type": "Microsoft.App/containerApps"
}
PS /home>

Figure 8.3 – Container app deployed

That's a lot of important JSON; one part will even give you the web address of the newly deployed containers. You can either scroll up to find it, use the Azure portal, or use the following command:

```
az containerapp show `
  --resource-group rg-containerapps `
  --name tightlycoupledcontainers `
  --query properties.configuration.ingress.fqdn
```

After a short while, you will see some output similar to the following screenshot:

```
PS /home> az containerapp show `
>>   --resource-group rg-containerapps `
>>   --name tightlycoupledcontainers `
>>   --query properties.configuration.ingress.fqdn
"tightlycoupledcontainers.ashywater-361bcfb8.westeurope.azurecontainerapps.io"
PS /home> █
```

Figure 8.4 – Web address for our deployed container

If we navigate to this web address, you will be presented with the **Cloud Adventures** online shop:

Production
V1
Cloud
Adventures

Figure 8.5 – Cloud Adventures online shop web page

Now that we have deployed and can access our containers successfully, it's time to think about scaling. Let's say our customers absolutely love the API and are starting to integrate it with their backend systems to automate the billing and invoicing process. Instead of just calling the API when they use the frontend, they are not continuously polling the API for any updates. We could have an opinion about that behavior but it is what customers do when you provide them with an API. In order to deal with that traffic, we need to allocate more resources to the API, not only to provide more *juice* but also in terms of availability. Even if a container starts very quickly, if it's being used often, we cannot afford it to be done for just a second. We want to run an additional replica of that container. But here's the catch – we want to do that for just that one container, not both.

In the preceding example, we would have to scale up the complete container app (which contains both containers). We are essentially adding resources to the frontend as well, even though the frontend doesn't need them. What we want is to independently scale both containers. Let's take a look at how we can achieve that!

In the next section, we will be using AKS to show how we can deploy and scale two containers independently from each other. We can use ACA for this as well but there are technical constraints within ACA that prevent us from doing exactly what we want. In the world of containers, this is always the trade-off: how much complexity are you willing to accept to take away constraints?

Loosely coupled

Before we start looking at deploying the two containers, we first need a resource group and an AKS cluster. If you need a recap about AKS, you can read *Chapter 6*. To create the resource group, use the following command:

```
az group create `
  --name rg-aks `
  --location westeurope
```

Within an instance, you should have some JSON output similar to the following screenshot to confirm that the resource group has been created successfully:

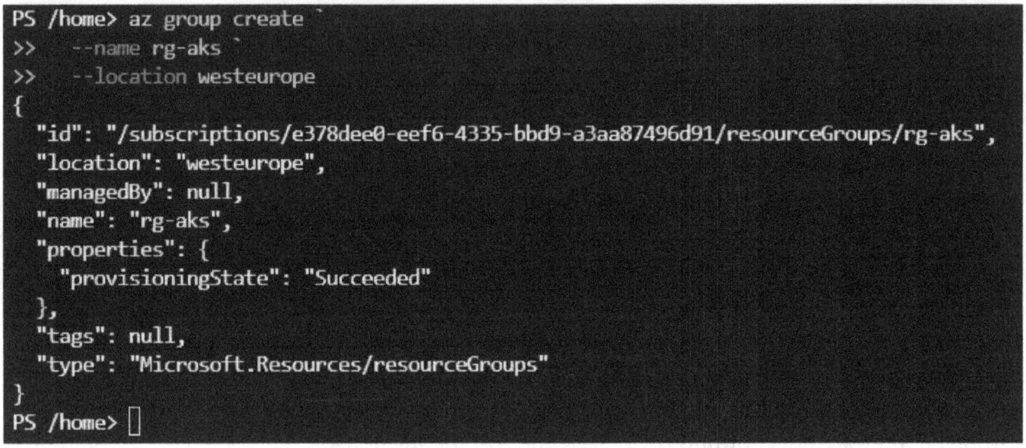
PS /home> az group create `
>> --name rg-aks `
>> --location westeurope
{
"id": "/subscriptions/e378dee0-eef6-4335-bbd9-a3aa87496d91/resourceGroups/rg-aks",
"location": "westeurope",
"managedBy": null,
"name": "rg-aks",
"properties": {
"provisioningState": "Succeeded"
},
"tags": null,
"type": "Microsoft.Resources/resourceGroups"
}
PS /home>

Figure 8.6 – Resource group created successfully

Now that we have the resource group, let's go ahead and create an AKS cluster. This cluster is going to be a very basic cluster and in no way fit for production use, but it is good enough for this example:

```
Az aks create `
--location westeurope `
--resource-group rg-aks `
--name aks `
--generate-ssh-keys
```

As you may already know if you read *Chapter 6*, this will create an AKS cluster in the `westeurope` Azure region, store it in the `rg-aks` resource group, and call it `aks`. It can take some time for the AKS cluster to be created, but once finished, you should have some output similar to the following screenshot:

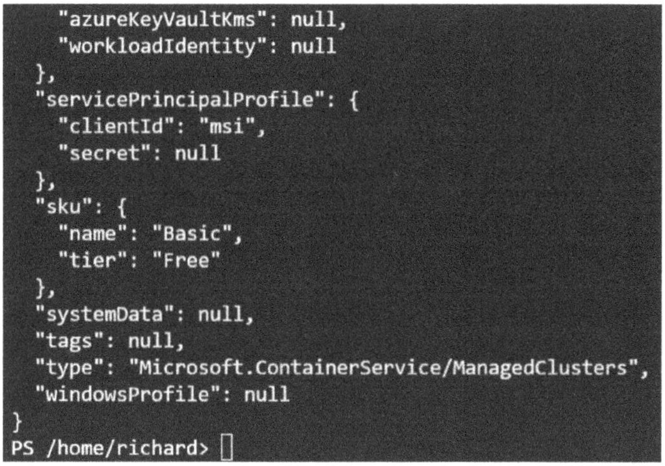
"azureKeyVaultKms": null,
"workloadIdentity": null
},
"servicePrincipalProfile": {
"clientId": "msi",
"secret": null
},
"sku": {
"name": "Basic",
"tier": "Free"
},
"systemData": null,
"tags": null,
"type": "Microsoft.ContainerService/ManagedClusters",
"windowsProfile": null
}
PS /home/richard>

Figure 8.7 – AKS cluster created

Now that the cluster is deployed, it is time to connect to it. For this, we will use the `az aks` command:

```
az aks get-credentials --resource-group rg-aks --name aks
```

By running this command, you will download and merge your cluster login details (context) to the `.kube/config` file in your profile:

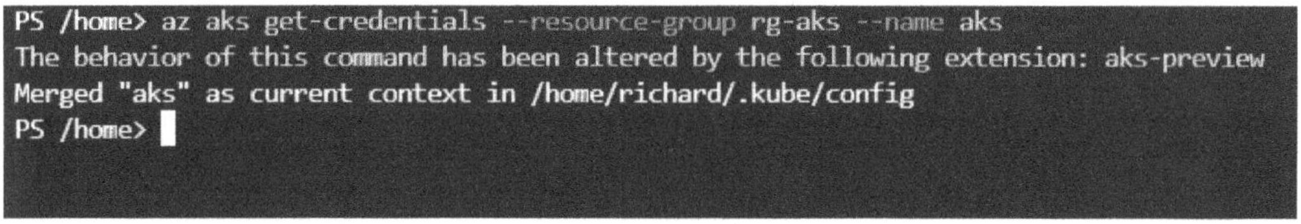
PS /home> az aks get-credentials --resource-group rg-aks --name aks
The behavior of this command has been altered by the following extension: aks-preview
Merged "aks" as current context in /home/richard/.kube/config
PS /home>

Figure 8.8 – Retrieving kubeconfig

With the cluster created and our cluster login details stored in our `.kubeconfig` file, it's time to start deploying our Kubernetes objects. As mentioned in the *Understanding how Azure Kubernetes Services work* section in *Chapter 6*, we are going to be creating Deployments on the cluster. In fact, we will be creating two Deployments, which will in turn create two Pods, one per Deployment. We will then also create two Service objects to allow the frontend and API to be accessible over the internet.

In *Chapter 6*, we mentioned that we will go deeper into Ingress. This is where we start. First, we need to install an Ingress controller. For this, we will use `ingress-nginx`.

> **Important note**
>
> Helm is a package manager for Kubernetes and is widely used to install applications. We will be using Helm to install the `ingress-nginx` Helm chart. Helm comes already installed with Cloud Shell. If you need to install it locally, please check out the Helm website for installation instructions: `https://helm.sh/docs/intro/install/`.

You can use the following commands to deploy `ingress-nginx`:

```
$NAMESPACE="ingress-basic"

helm repo add ingress-nginx https://kubernetes.github.io/
ingress-nginx
helm repo update

helm install ingress-nginx ingress-nginx/ingress-nginx `
  --create-namespace `
  --namespace $NAMESPACE `
  --set controller.service.annotations."service\.beta\.
kubernetes\.io/azure-load-balancer-health-probe-request-path"=/
healthz
```

After the nginx Ingress has been installed successfully, you will see some help text on how to create an Ingress object, as shown in the following screenshot:

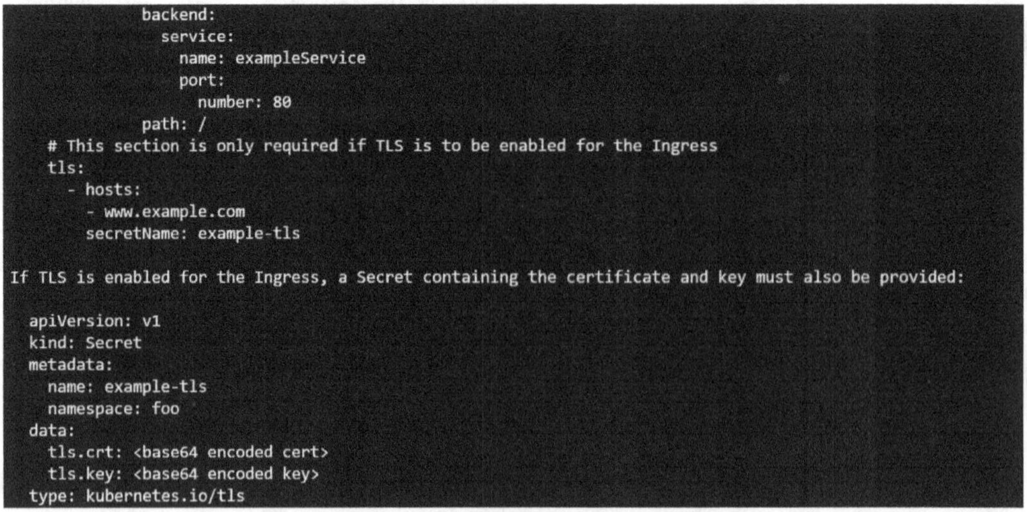

backend:
service:
name: exampleService
port:
number: 80
path: /
This section is only required if TLS is to be enabled for the Ingress
tls:
- hosts:
- www.example.com
secretName: example-tls
If TLS is enabled for the Ingress, a Secret containing the certificate and key must also be provided:
apiVersion: v1
kind: Secret
metadata:
name: example-tls
namespace: foo
data:
tls.crt: <base64 encoded cert>
tls.key: <base64 encoded key>
type: kubernetes.io/tls

Figure 8.9 – nginx Ingress installed

This is super helpful and has been created by the maintainers of the Helm chart. So, if you are installing another Helm chart, you might not have such helpful text.

To confirm that the nginx Ingress has been installed, use the following command:

```
kubectl get pods -n ingress-basic
```

If you see `1/1 READY`, then you are all good.

Now, it's time to deploy our apps. Let's start with the frontend Deployment, Service, and Ingress.

In the GitHub repo under the `CH08` folder, you will find a file called `frontend.yaml`. The file may look like a lot but we explained it back in *Chapter 6* in the *Deploying containers to AKS* section, so if you have not read that chapter, we would highly recommend it.

Once you're all caught up on AKS, it's now time to apply this manifest file to the cluster. As it is all stored in the GitHub repo, you don't need to download it locally; you can just use the following command and it will apply the manifest directly from the repo:

```
kubectl apply -f https://raw.githubusercontent.com/
PacktPublishing/-Up-and-Running-with-Azure-Containers/main/
CH08/frontend.yaml
```

Within a second or two, you should have some output as in the following screenshot:

PS /home> kubectl apply -f https://raw.githubusercontent.com/PacktPublishing/-Up-and-Running-with-Azure-Containers/main/CH08/frontend.yaml
deployment.apps/frontend created
service/frontend created
ingress.networking.k8s.io/frontend created
PS /home>

Figure 8.10 – Applying the frontend YAML configuration using kubectl

With the billing API deployed, it's time to add the `billingstatement` API Deployment and Service. Again, we have the manifest file stored in the repo for you. All you have to do is run the following command:

```
kubectl apply -f https://raw.githubusercontent.com/
PacktPublishing/-Up-and-Running-with-Azure-Containers/main/
CH08/api.yaml
```

Once applied, you should have output similar to the following screenshot:

PS /home> kubectl apply -f https://raw.githubusercontent.com/PacktPublishing/-Up-and-Running-with-Azure-Containers/main/CH08/api.yaml
deployment.apps/billingstatements created
service/billingstatements created
ingress.networking.k8s.io/frontend configured
PS /home>

Figure 8.11 – Applying the API YAML configuration using kubectl

Now that we have deployed both the frontend and the API, we should verify that they are up and running and also the IP address they are listening on.

> **Important note**
>
> Normally, you would not use the `loadBalancer` Service type to expose your application to the world. It would normally be behind an Ingress controller, but for the sake of this example, we will forgo the Ingress controller.

To do this, you can use the `kubectl get` command:

```
kubectl get deployments,pods,service,ingress
```

You should now see the output of the Deployments, Pods, Services, and Ingress Kubernetes objects. This should look something like the following screenshot:

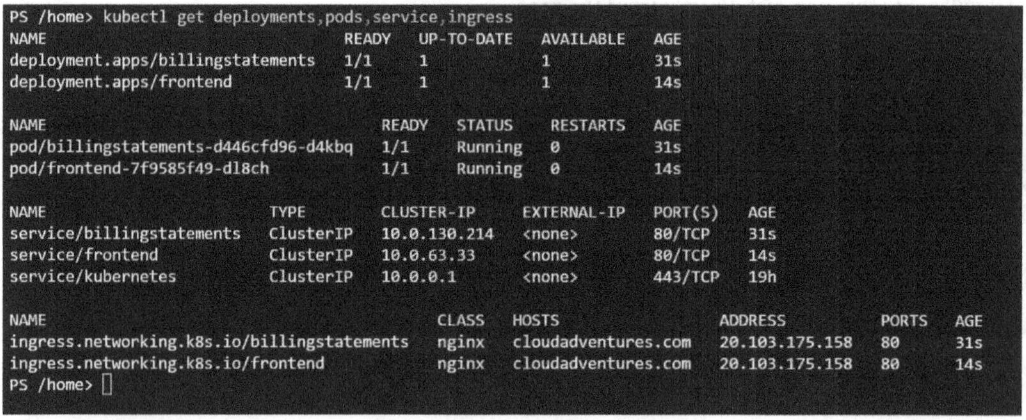

```
PS /home> kubectl get deployments,pods,service,ingress
NAME                                 READY   UP-TO-DATE   AVAILABLE   AGE
deployment.apps/billingstatements    1/1     1            1           31s
deployment.apps/frontend             1/1     1            1           14s

NAME                                     READY   STATUS    RESTARTS   AGE
pod/billingstatements-d446cfd96-d4kbq    1/1     Running   0          31s
pod/frontend-7f9585f49-dl8ch             1/1     Running   0          14s

NAME                         TYPE        CLUSTER-IP     EXTERNAL-IP   PORT(S)   AGE
service/billingstatements    ClusterIP   10.0.130.214   <none>        80/TCP    31s
service/frontend             ClusterIP   10.0.63.33     <none>        80/TCP    14s
service/kubernetes           ClusterIP   10.0.0.1       <none>        443/TCP   19h

NAME                                            CLASS   HOSTS                 ADDRESS          PORTS   AGE
ingress.networking.k8s.io/billingstatements     nginx   cloudadventures.com   20.103.175.158   80      31s
ingress.networking.k8s.io/frontend              nginx   cloudadventures.com   20.103.175.158   80      14s
PS /home>
```

Figure 8.12 – kubectl get output

You will see that the Ingress object has a column called `HOSTS` and a column called `ADDRESS`. The `HOSTS` value is supplied via the YAML manifest from the GitHub repo. The `ADDRESS` value is an IP address from Azure attached to a load balancer. In this case, they are the same, but you may have different ones if your application requires them.

If you have used the hostname provided, you will need to update your local host file on your computer with the `HOSTS` and `ADDRESS` details to allow you to view the website. If you have changed the `HOSTS` value to a domain you own, you can just update your DNS entry. Updating your DNS and host file is out of scope, but a simple Google search will help you to update your DNS or host file.

If you now browse to the domain name linked to the Ingress IP address (for me, it is `cloudadventures.com`), you should see the Cloud Adventures webshop, as in the following screenshot:

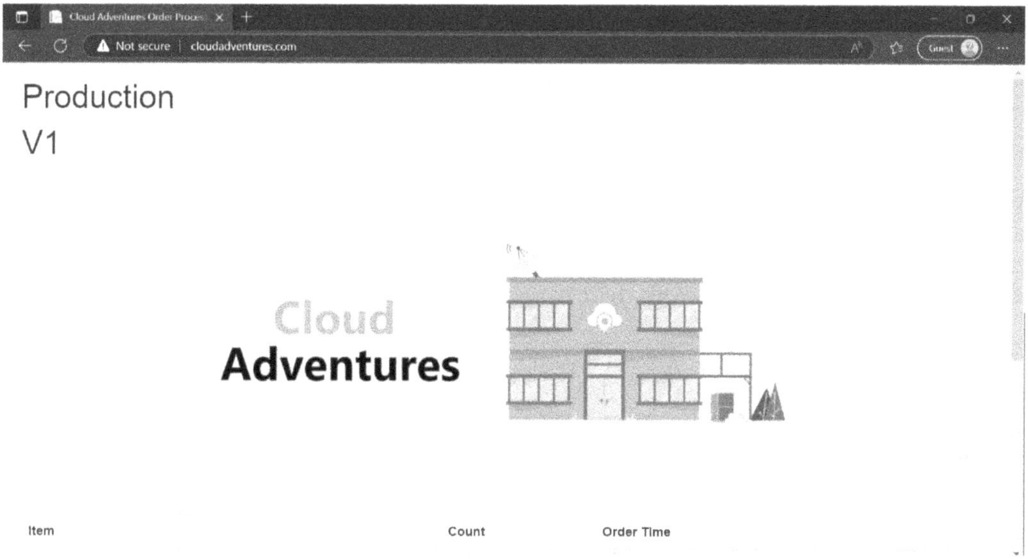
Not secure
cloudadventures.com
Production
V1
Cloud
Adventures
Item
Count
Order Time

Figure 8.13 – Cloud Adventures website

Let's check whether the API is up and running. We can use the same domain name as before; just add `/billingstatement` to the end and you should be able to see the API results:

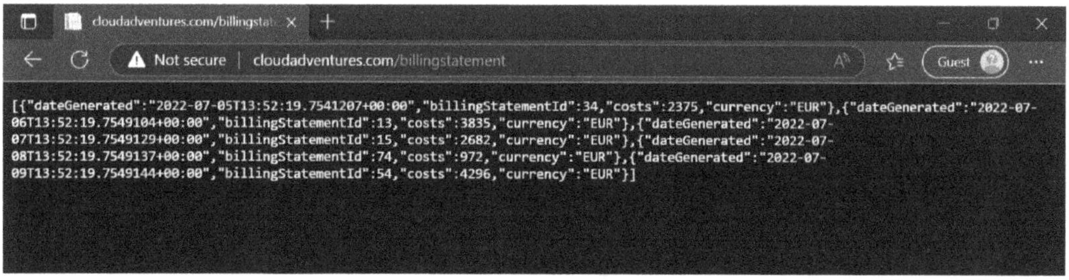
Not secure
cloudadventures.com/billingstatement
[{"dateGenerated":"2022-07-05T13:52:19.7541207+00:00","billingStatementId":34,"costs":2375,"currency":"EUR"},{"dateGenerated":"2022-07-06T13:52:19.7549104+00:00","billingStatementId":13,"costs":3835,"currency":"EUR"},{"dateGenerated":"2022-07-07T13:52:19.7549129+00:00","billingStatementId":15,"costs":2682,"currency":"EUR"},{"dateGenerated":"2022-07-08T13:52:19.7549137+00:00","billingStatementId":74,"costs":972,"currency":"EUR"},{"dateGenerated":"2022-07-09T13:52:19.7549144+00:00","billingStatementId":54,"costs":4296,"currency":"EUR"}]

Figure 8.14 – The billingstatement API website

Now that we know we can reach both sites, let's think about scaling as we did in the *Tightly coupled* section. We will use the same scenario where the customers are loving the API and using it more than the frontend. Earlier, when we needed to scale the API, we had to scale both the frontend and the API together as they were tightly coupled together in one container app.

With the applications running in Kubernetes, we can independently scale each application as they are running in their own Deployment. As we are using a Service object also, we do not need another public IP address for the new API instances as the Service handles this for us internally on AKS.

Let's look at adding another instance, or *replica* as it is known in Kubernetes, of the API. To do this, we will use the following command:

```
kubectl scale deployments billingstatements --replicas=3
```

> **Important note**
>
> Although it is nice and easy to use the `kubectl` command to scale a Deployment, it is highly recommended to change the replica count in the manifest to the amount you need and apply the manifest again. You would ideally keep the manifests in a GitHub repo and only apply what is in your deployment branch. This will help ensure your cluster does not have ad hoc changes, and you have an audit trail of who made what changes and when.

What this command does is tell the Kubernetes API that we want three replicas of the `billingstatements` API running. So, in this example, we will be going from one to three replicas. We can use the `kubectl get` command from earlier to scale the billing API Deployment to have three replicas:

```
PS /home> kubectl scale deployments billingstatements --replicas=3
deployment.apps/billingstatements scaled
PS /home> kubectl get deployment billingstatements
NAME                READY   UP-TO-DATE   AVAILABLE   AGE
billingstatements   3/3     3            3           7m52s
PS /home>
```

Figure 8.15 – The billingstatements deployment scaled

You will notice that `deployment.apps/billingstatement` is now showing `3/3` in the `READY` column and `deployment.apps/frontend` is still showing `1/1`.

This means we are now only allocating resources to those Services that need them and not wasting any money on scaling the frontend, which does not need scaling.

We don't want to end up with an invoice for unused resources at the end of this month so let's delete the resources you created during this chapter. We can do this by deleting the resource group. You can do this by using the following commands:

```
az group delete --name "rg-containerapps"
az group delete --name "rg-aks"
```

With that, you have now deployed both scenarios, tested whether they are working, and removed the Azure resources from your subscription to ensure you do not rack up a big bill. Let's summarize what we've seen so far.

Summary

In this chapter, we learned about the different ways infrastructure behaves if we deploy a tightly or loosely coupled solution. In the examples, we showed the technical impact, which teaches us that we really need to think about this not only from a technical perspective but maybe even more from a capacity perspective. Simply scaling everything will result in quite the bill and compute resource inefficiency, something that is exactly the opposite of what containers should be – efficient and cost-effective.

We also learned that running Windows containers is perfectly acceptable given the correct use case. It is perfect for scenarios where we cannot simply allocate thousands of hours for development but need to spread the workload.

In the next chapter, we will continue with the e-commerce shop examples and gradually add more features as our fictional company progresses.

9
Container Technologies for Startups

We've seen single container deployments and multiple container deployments, and now it's time to bring containers to the real world. You're a start-up, starting from scratch, and you are building a solution on container technology. What should you do?

In the previous chapters, we have gone through the technical side of decision-making. Whether you are coming from a greenfield scenario or a scenario where you have to migrate and refactor existing code, there is always a scenario for you.

But now you're a start-up. You have some plans (usually including a business plan) and somewhat of a technical direction that you want to move in. So let's say that the technical direction is containers. In such a case, it's time to take a look at real-world implementations.

You could argue "*Does it matter if I'm a start-up or a medium to small business?*" Yes, it does matter. There are considerable differences in customer count, finances, and even employee knowledge. It doesn't mean it's the worst place to start.

Your technical success depends on one thing: planning. What decisions can you make now that will benefit you in the future? Containerization is one of those things. Even if the underlying technology stack/infrastructure is different, a container will always be a container.

In this chapter, we're going to go over the following topics:

- A common start-up scenario and solution design
- Deploying and implementing the solution
- Learnings and future decisions

A common start-up scenario and solution design

Designing a solution is often overlooked and sometimes even underestimated. A design is not just about the diagram, it is about the process that has to be followed to get there. It helps us think things through and the result will help us convince internal stakeholders to answer the question "*Why did we make these decisions?*"

In this section, we will break down a use case and translate the scenario into a series of technical and business requirements. Once we have the requirements clear, we will be able to make technological decisions and choose what technology is the best fit for our use case. Finally, we will visualize the design so that we can present it internally.

Let's start with a common scenario. Keep in mind that there are many scenarios, but the approach will remain the same.

An e-commerce company called Cloud Adventures provides a webshop for its customers that allows them to buy Office 365 licenses. The solution consists of a frontend portal and a backend API that processes the billing. Additionally, the backend API allows direct access by the customer to automate billing processes and integrate with their existing **customer relationship management** (**CRM**) system.

The Cloud Adventures customers are **Cloud Solution Providers** (**CSP**). As the solution is connected to Microsoft Partner Center, they require absolute isolation and a single-tenant deployment. In no way should the API of one customer be able to access the API of another customer.

As CSPs can have many hundreds of end customers, the solution needs to be scalable. From a software architecture perspective, the solution is stateless and can be restarted and scaled at any time.

The solution is written in .NET 6 and passes all tests for containerization. From a technical perspective, the developers require several environment variables to be passed:

- >>Billing API address
- >>Billing API key
- >>Azure SQL connection string

As security is high on the list, they also require network traffic to be isolated within a virtual network.

Cloud Adventures has asked you to design and deploy an infrastructure that will help their developers deploy the solution to Microsoft Azure container technologies.

It's important to know that Cloud Adventures currently is training their Ops personnel on the job. As the current focus is launching the first couple of customers, the infrastructure may not require extensive management or knowledge from the Ops team. From a financial perspective, costs need to be kept to a minimum.

That's quite the use case! But let's dissect it before we start designing a solution. What do we know?

- **Security**: From a security perspective, we know a couple of things:
  - Isolation between customers is a hard requirement; the solution must be deployed as *single-tenant*. This means deploying a combination of containers for each customer.
  - Cloud Adventures requires integration with virtual networks to keep the traffic isolated.
- **Management**: From a management perspective, we have some additional facts:
  - The Ops team is still learning on the job, so complexity must be kept to a minimum.
  - The current focus is on launching customers on the platform. Other than knowledge, time and priority are constraints for spending time on managing the infrastructure.
  - The solution must be scalable.
- **Financial**: We really only have one requirement here. Costs need to be kept to a minimum.
- **Technical**: From a merely technical perspective, we have some other requirements to deal with:
  - There is a frontend that requires access. We will assume this happens over HTTPS.
  - The Billing API requires direct access for customers to connect to and integrate with their CRM.
  - Environment variables must be passed upon runtime.

Choosing the right technology

This is probably the hardest thing to do. When choosing the technology stack, we have to take all requirements into account. Let's just put all technologies into a single table and see whether we can provide motivation as to whether to use them or not.

Technology	Yes/No	Motivation
Azure Functions for Containers	No	We want to host a frontend and an API per customer. Azure Functions has a completely different use case. However, if it was just for the API, this could be an option.
Azure App Services for Containers	No	Running multiple containers that are scalable requires different containers on different app services or would require Docker Compose, which makes the containers tightly coupled from a scaling perspective.

Technology	Yes/No	Motivation
Azure Container Instances	No	Even though ACI provides absolute isolation, it is fairly limited when it comes to scaling. Serving hundreds of customers would potentially hit the limits of the container groups.
Azure Container Apps	Yes	Azure Container Apps supports virtual network integration and is essentially an automagically managed Azure Kubernetes Service. We can provide the Ops team with a lot of options without the management burden. Additionally, providing a set of container apps per customer also helps with scaling per customer and determining the infrastructure costs for customers.
Azure Kubernetes Services	No	Technically yes, but the management overhead/required knowledge might be too much for the Ops team. If the solution proves to be successful, this might be the next step.

Table 9.1 – Which technology fits

So, we kind of decided that **Azure Container Apps** (**ACA**) is going to be the go-to technology for Cloud Adventures as a start-up. It's important to understand that ACA provides more than just peace of mind for the Ops team. As ACA tightly integrates with features such as Dapr and KEDA, which we have discussed before, we are also providing the developers with a world of options. If you haven't done so, please check *Chapter 5*.

Okay, so ACA it is. Let's visualize this.

Visualizing a design

Visualizing a design helps both the people who manage it, the developers, and the customer if they require documentation. Let's see what these requirements will look like in a visualization. We could go into a lot of detail, but let's go with something not too generic and not too detailed instead:

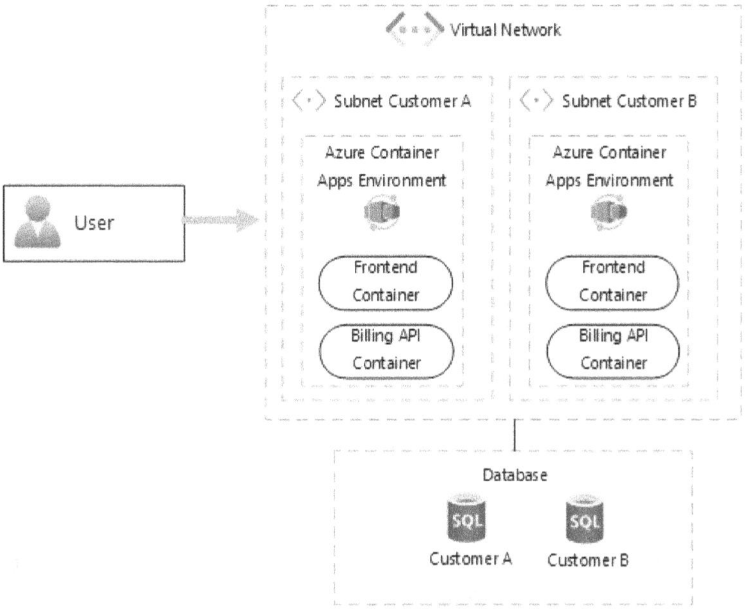
Virtual Network
Subnet Customer A
Subnet Customer B
User
Azure Container Apps Environment
Azure Container Apps Environment
Frontend Container
Billing API Container
Frontend Container
Billing API Container
Database
SQL
SQL
Customer A
Customer B

Figure 9.1 – Azure Container Instances design

What we visualized in this design is that we are going to build something that provides isolation between customers, virtual network integration, and a connection to a database. Please note that this doesn't mean all other requirements are out the window. However, we did decide on Azure Container Apps specifically because it fulfills a lot of those requirements already.

> **Important note**
>
> In the design, you see Azure SQL as the database solution. Azure SQL supports **virtual network** (**vNet**) service endpoints, which allow for a connection over the virtual network to the container of a customer.

You might be thinking, "*So how does this diagram help?*" Well, it's not only good to visualize your approach, but it's also a form of documentation. What you will see in the next chapters is that we will transform this design as the Cloud Adventures organization becomes more successful. Keeping track and continuously updating your documentation ensures that *everyone* in your organization understands what's going on. Believe us, you will regret it if you don't! We've been there.

Deploying and implementing the solution

Now it's time to deploy the solution. What we need is the following:

- ACA with vNet integration
- Two containers (frontend and Billing API)
- Isolation per customer (single tenancy)
- Environment variables (Billing API address, Billing API key, and Azure SQL connection string)

You could argue that you could use much more, and that's true. But remember: Cloud Adventures is a start-up, and we want to start small. Once we gain financial momentum, we will start making big leaps!

> **Important note**
>
> In the design (*Figure 9.1*), we are leveraging Azure SQL as a database solution. Deploying and managing Azure SQL is beyond the scope of this book. The examples provided are to show you have to pass environment variables such as connection strings to connect your solution to a database, as you will not come across many real-life scenarios that do not use any sort of database.

Let's look at what it takes to deploy the solution using the Azure CLI.

Deploying the resource group and networking

As always, we first need to start with a resource group. You can use the following command to create one:

```
az group create `
  --name rg-cloud-adventures `
  --location westeurope
```

You will instantly have some output similar to what's shown in *Figure 9.2*:

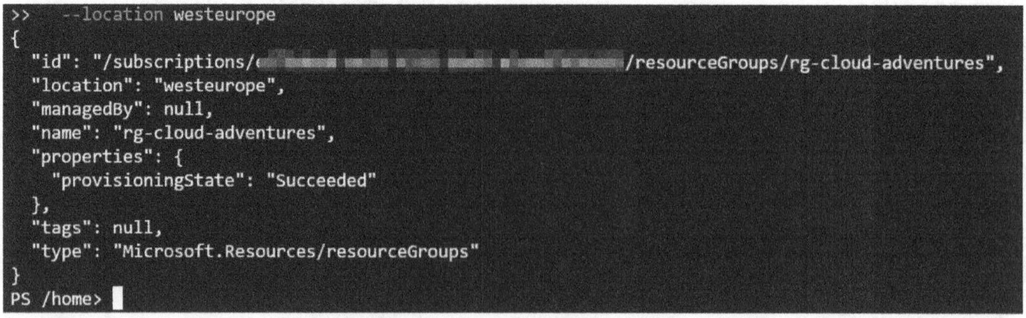
>>   --location westeurope
{
"id": "/subscriptions/[redacted]/resourceGroups/rg-cloud-adventures",
"location": "westeurope",
"managedBy": null,
"name": "rg-cloud-adventures",
"properties": {
"provisioningState": "Succeeded"
},
"tags": null,
"type": "Microsoft.Resources/resourceGroups"
}
PS /home>

Figure 9.2 – Resource group created

Now that we have the resource group, we need to create the vNet that we will be using to host each customer's subnet. To do this, use the following command:

```
az network vnet create `
--resource-group rg-cloud-adventures `
--name vnet-cloud-adventures `
--location westeurope `
--address-prefix 10.0.0.0/16
```

After a short while, you should see some output similar to *Figure 9.3*:

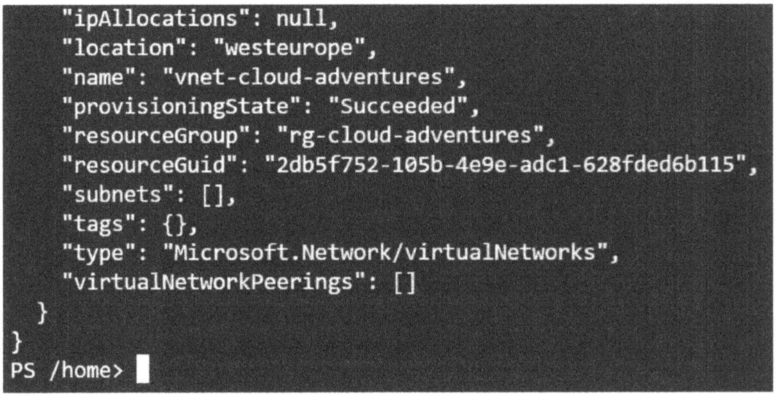
"ipAllocations": null,
"location": "westeurope",
"name": "vnet-cloud-adventures",
"provisioningState": "Succeeded",
"resourceGroup": "rg-cloud-adventures",
"resourceGuid": "2db5f752-105b-4e9e-adc1-628fded6b115",
"subnets": [],
"tags": {},
"type": "Microsoft.Network/virtualNetworks",
"virtualNetworkPeerings": []
}
}
PS /home>

Figure 9.3 – vNet created

Then we need to create the customer subnets, for which we will use the following command:

```
az network vnet subnet create `
--resource-group rg-cloud-adventures `
--vnet-name vnet-cloud-adventures `
--name customer-a `
--address-prefixes 10.0.0.0/23
```

Once the command has been completed you will see some output as in *Figure 9.4*:

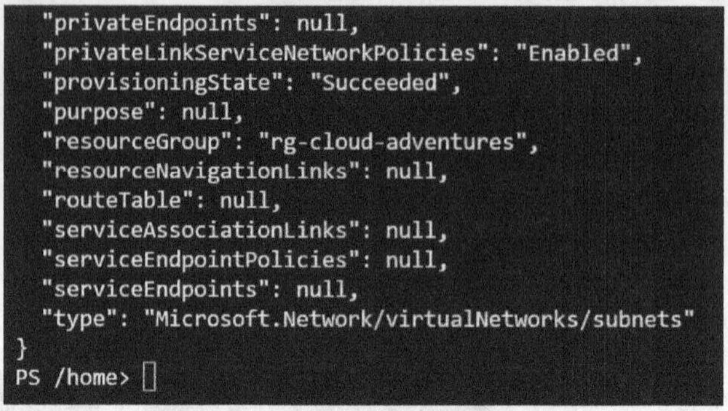
"privateEndpoints": null,
"privateLinkServiceNetworkPolicies": "Enabled",
"provisioningState": "Succeeded",
"purpose": null,
"resourceGroup": "rg-cloud-adventures",
"resourceNavigationLinks": null,
"routeTable": null,
"serviceAssociationLinks": null,
"serviceEndpointPolicies": null,
"serviceEndpoints": null,
"type": "Microsoft.Network/virtualNetworks/subnets"
}
PS /home>

Figure 9.4 – vNet subnet created

Next, you need to find the resource ID of the newly created subnet. This will be needed when creating the container app environment. You can either scroll up in the output to find it or use the following command:

```
$subnetId = az network vnet subnet show `
--resource-group rg-cloud-adventures `
--vnet-name vnet-cloud-adventures `
--name customer-a `
--query "id" -o tsv | tr -d '[:space:]'
```

You should now see the subnet ID as in *Figure 9.5*:

PS /home> $subnetId = az network vnet subnet show `
>> --resource-group rg-cloud-adventures `
>> --vnet-name vnet-cloud-adventures `
>> --name customer-a `
>> --query "id" -o tsv | tr -d '[:space:]'
PS /home> $subnetId
/subscriptions/e[redacted]/resourceGroups/rg-cloud-adventures/providers/Microsoft.Network/virtualNetworks/vnet-cloud-adv
entures/subnets/customer-a
PS /home>

Figure 9.5 – Getting the subnet resource ID

Now it's time to get to the fun part: deploying the container app environment.

Deploying the customer's container app environment

To deploy the per-customer container app environment, you can use the following command:

```
az containerapp env create `
--name ace-customer-a `
--resource-group rg-cloud-adventures `
--location westeurope `
--infrastructure-subnet-resource-id $subnetId
```

After a few seconds, you should see the same output as in *Figure 9.6*. This tells us the container app environment has been created:

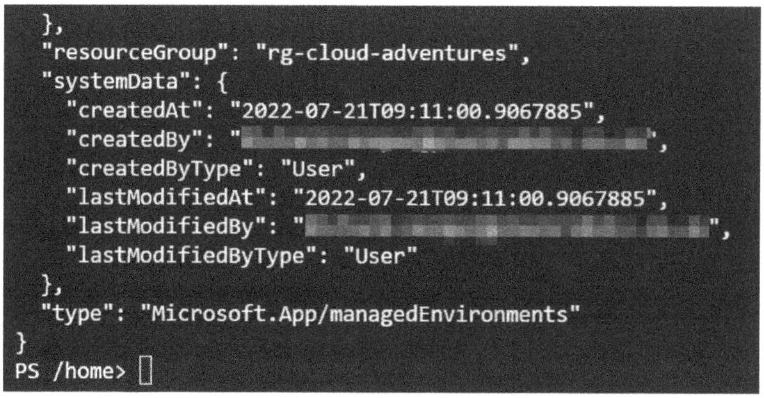
},
"resourceGroup": "rg-cloud-adventures",
"systemData": {
"createdAt": "2022-07-21T09:11:00.9067885",
"createdBy": "[redacted]",
"createdByType": "User",
"lastModifiedAt": "2022-07-21T09:11:00.9067885",
"lastModifiedBy": "[redacted]",
"lastModifiedByType": "User"
},
"type": "Microsoft.App/managedEnvironments"
}
PS /home>

Figure 9.6 – Container app environment created

> **Important note**
>
> If you haven't run the preceding command to get the subnet resource ID, and instead have just copied it into your clipboard, replace `$subnetID` with the actual ID.

It's now time to deploy and test the containers.

Deploying and testing the containers

We will start with the API container as we need some information from it. This information will be used by the frontend container as environment variables.

To create the API container, we will use the following command:

```
az containerapp create `
--name customer-a-api `
--resource-group rg-cloud-adventures `
--environment ace-customer-a `
--image whaakman/container-demos:billingstatementsv3 `
--ingress external `
--target-port 80 `
--query properties.configuration.ingress.fqdn
```

Depending on how big the container is, a few seconds or minutes later, you will have some output similar to the following screenshot:

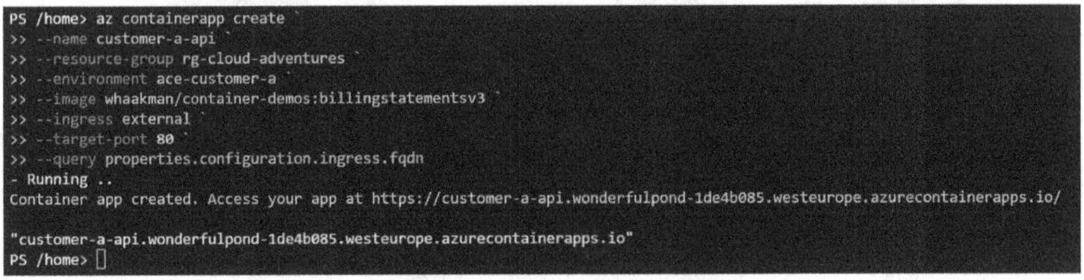
PS /home> az containerapp create `
>> --name customer-a-api `
>> --resource-group rg-cloud-adventures `
>> --environment ace-customer-a `
>> --image whaakman/container-demos:billingstatementsv3 `
>> --ingress external `
>> --target-port 80 `
>> --query properties.configuration.ingress.fqdn
- Running ..
Container app created. Access your app at https://customer-a-api.wonderfulpond-1de4b085.westeurope.azurecontainerapps.io/
"customer-a-api.wonderfulpond-1de4b085.westeurope.azurecontainerapps.io"
PS /home>

Figure 9.7 – API container deployed

With that, the backend is deployed and, thanks to the `--query` parameter, we know the API's public address. Navigate to the site to check whether it is up and running. As it is the Billing API, we need to add `/billingstatement` to the end of the web address, as can be seen in the following screenshot:

https://customer-a-api.wonderfulpond-1de4b085.westeurope.azurecontainerapps.io/billingstatement
Guest
[{"dateGenerated":"2022-07-22T10:51:41.1987616+00:00","billingStatementId":61,"costs":541,"currency":"EUR"},{"dateGenerated":"2022-07-23T10:51:41.1995539+00:00","billingStatementId":64,"costs":4753,"currency":"EUR"},{"dateGenerated":"2022-07-24T10:51:41.199556+00:00","billingStatementId":17,"costs":4340,"currency":"EUR"},{"dateGenerated":"2022-07-25T10:51:41.1995567+00:00","billingStatementId":86,"costs":827,"currency":"EUR"},{"dateGenerated":"2022-07-26T10:51:41.1995574+00:00","billingStatementId":85,"costs":4546,"currency":"EUR"}]

Figure 9.8 – API container website success

This web address is helpful to give to your customers, but also for the frontend container, as it will need to know how to reach the API. So, take note of this for the next step, which is to deploy the frontend. The command for this is very similar to the backend command except that we change the container image and also add some environment variables.

When we look at the scenario, the three environment variables we need to supply to the frontend container are the Billing API address, the Billing API key, and the Azure SQL connection string. For the Billing API key and the Azure SQL connection string, we are going to use some made-up values, but you would normally know these from your deployed SQL server and, for the key, your own tooling. For the Billing API address, we have the web address from before, but this is the external web address. We don't really want the frontend going out of the ACA environment to go back in. So, what we can do is add the word `internal` just after the app name in the web address. We currently have the following:

```
https://customer-a-api.wonderfulpond-1de4b085.westeurope.
azurecontainerapps.io
```

For internal usage, it will look like this:

```
https://customer-a-api.internal.wonderfulpond-1de4b085.westeurope.
azurecontainerapps.io
```

With this new web address, we are ready to deploy the frontend container app. To do this, we will use the following command:

```
az containerapp create `
--name customer-a-frontend `
--resource-group rg-cloud-adventures `
--environment ace-customer-a `
--image whaakman/container-demos:cloudadventuresshopv1 `
--ingress external `
--target-port 80 `
--env-vars BILLING_API_ADDRESS="https://customer-a-api.
internal.wonderfulpond-1de4b085.westeurope.azurecontainerapps.
io" BILLING_API_KEY="API key" SQL_CON_STRING="Azure sql
connection string" `
--query properties.configuration.ingress.fqdn
```

After a short while, you should see the web address of the container app as in *Figure 9.9*. This informs us that the command completed:

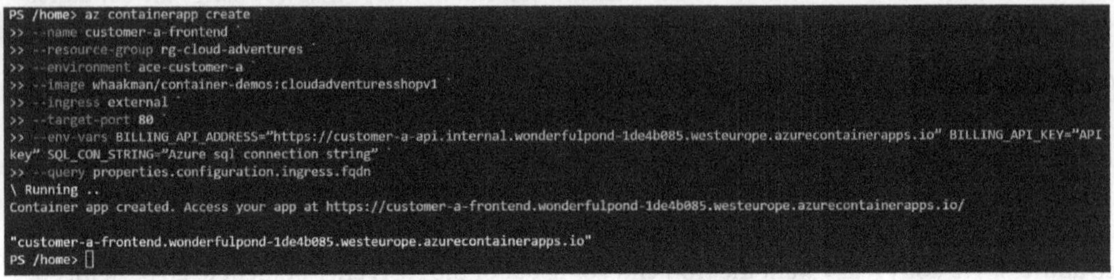

```
PS /home> az containerapp create `
>> --name customer-a-frontend `
>> --resource-group rg-cloud-adventures `
>> --environment ace-customer-a `
>> --image whaakman/container-demos:cloudadventuresshopv1 `
>> --ingress external `
>> --target-port 80 `
>> --env-vars BILLING_API_ADDRESS="https://customer-a-api.internal.wonderfulpond-1de4b085.westeurope.azurecontainerapps.io" BILLING_API_KEY="API key" SQL_CON_STRING="Azure sql connection string" `
>> --query properties.configuration.ingress.fqdn
\ Running ..
Container app created. Access your app at https://customer-a-frontend.wonderfulpond-1de4b085.westeurope.azurecontainerapps.io/

"customer-a-frontend.wonderfulpond-1de4b085.westeurope.azurecontainerapps.io"
PS /home> 
```

Figure 9.9 – Frontend container deployed

That's the frontend deployed, and as we used the `--query` parameter again, we know the frontend's public address. Navigate to the site using your web browser to check whether it is up and running:

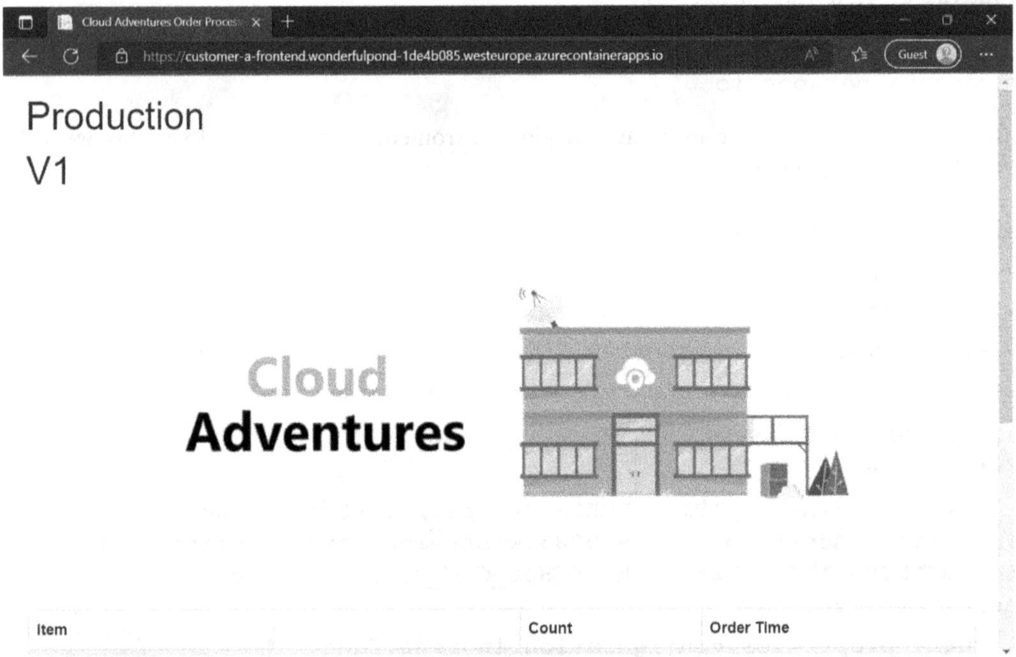
Cloud Adventures Order Proces
https://customer-a-frontend.wonderfulpond-1de4b085.westeurope.azurecontainerapps.io
Guest
Production
V1
Cloud
Adventures
Item
Count
Order Time

Figure 9.10 – Frontend container running

With that, customer A is all set up and ready to go. To add customer B, we will need to repeat some steps. Don't worry, I won't copy and paste the same steps here. But, if you do want to follow along and add customer B, you will need to start from the subnet creation part earlier in this chapter and follow along all the way till this point. Just make sure you change `customer-a` to `customer-b`.

> **Important note**
>
> Those of you with eagle eyes may have noticed that when we created the container apps environment, a Log Analytics workspace was created for us automatically. You can create one in advance and link it to the container environment by supplying the `--logs-workspace-id` and `--logs-workspace-key` parameters. In this scenario, it would be advisable to have one Log Analytics workspace per customer.

As you can see from the Deployments, we have managed to hit every point in the scenario. Now, if this was for a real customer, we would also add things such as custom domains, SSL certificates, and so on. They are outside the scope of this book, but if you are going to use ACA, I would highly recommend adding both custom domains and SSL certificates.

We don't want to end up with an invoice for unused resources at the end of this month, so let's delete the resources you created during this chapter. We can do this by deleting the resource group using the following command:

```
az group delete --name "rg-cloud-adventures"
```

Now that we have deployed the resources, let's look at what we have learned and any future decisions we may have to make.

Learnings and future decisions

We haven't really broken anything here, so the question is *how much was there to learn?* Well, a lot! We can deploy a production solution without using a million different technologies and with minimum complexity.

Specifically, what we've learned is that designing and deploying a solution is a very pragmatic but not very complex approach. Abstracting away the requirements from a use case and visualizing those requirements is very helpful.

There's also a cause for concern. Imagine, in this scenario, that we are a start-up; we don't really want to spend 100% of our time automating everything. We need to add value to the business and we need to do that quickly. ACA really helps us with that as it abstracts away a lot of the complexity that Kubernetes brings. But, as we've seen in the passing of environment variables, some things are not that flexible. Take the default hostname for example `https://customer-a-api.internal.wonderfulpond-1de4b085.westeurope.azurecontainerapps.io`.

A random name with a random number is not that great to automate against. To help you achieve automation, we have shown you how to use the `–query` parameter and you can go from there. However, in bigger enterprise scenarios, we probably want to have some level of control over those internal hostnames. Not only for automation purposes but for naming convention and governance as well. Nothing we need to worry about now, but we are making a mental note and will return to this subject in the upcoming chapters.

As for future decisions – yes, there are some to be made. The main question is *where do we go from here?* Well realistically, the only big next step from ACA is Azure Kubernetes Service. But it also depends on what your future requirements are. There will be a point in time when Cloud Adventures is successful and requires more and more Azure Container App environments. That will eventually turn out to be a costly endeavor, and Azure Kubernetes Service will turn out to be the more financially interesting technology.

That decision, however, depends on multiple factors:

- Do you still need to invest in knowledge for your Ops team?
- Is the complexity worth the costs in general?
- Can we build the isolation in Azure Kubernetes Service with at least the level of security we have in ACA?

There really are no right or wrong answers here. But in reality, we do see companies moving to Azure Kubernetes Service once they grow. The most important thing is to stick to the standards, and moving between technologies will not be a big issue. We will learn more about that in the upcoming chapters.

Summary

In this chapter, we have learned that a pragmatic approach to designing and implementing a solution doesn't have to be that complex. We started from the perspective of a start-up and, to be fair, nothing is as easy as deploying a greenfield solution as long as we stick to the requirements. Things will become more complicated once companies grow, business and technical requirements change, and we have to move between technologies.

We have also seen that integrating a solution with an Azure vNet is not that difficult. From an Ops perspective, it often feels complex but in reality, it usually just requires an additional one or more commands. In *Figure 9.1*, however, we can see that we are talking about different subnets. Luckily, Azure automates most of that for you, but you need to keep in mind the subnetting constraints before you run into the limits of your vNet.

In the next chapter, we will look at different topologies such as hub-spoke networking and vNet peering. Nothing to be afraid of; it's actually quite fun!

10
Container Technologies for Small and Medium-Sized Businesses

As a start-up, your goal is to build a minimum viable product, refine it, and make sure you can go to market as fast as possible. You usually don't have that many requirements from a customer. During the start-up phase of a business in the software industry, the focus is mainly on functionality.

But then you start becoming more successful; you're growing, you're scaling up, and soon you will be a small- or medium-sized business. Congratulations! However, this means you have a solid customer base and you have been servicing those customers for some time now. And, as it normally goes, customers want things from you! This can be a functionality that you are most likely already keeping track of, or something else. And that something else is usually what a lot of start-ups neglect, as the financial overhead or the investment is simply not worth it *yet*. We're talking about governance and security. Growing as a company also results in different (bigger) customers. Those customers will have requirements in the realm of governance and security.

That means refactoring your solution and infrastructure to keep up with the requirements. The bigger and more successful you become, the more of those requirements will arise. In this chapter, we will make a brief start by looking at a more scalable infrastructure that has the capabilities to fulfill those requirements.

In this chapter, we're going to go over the following topics:

- Cloud Adventures web shop – scenario and design
- Deploying and implementing a solution
- Learnings and future decisions

Cloud Adventures web shop – scenario and design

Let's continue and build on top of the scenario that we presented in the previous chapter.

An e-commerce company called Cloud Adventures provides a web shop for its customers that allows them to buy Office 365 licenses. The solution consists of a frontend portal and a backend API that processes the billing. Additionally, the backend API allows direct access by the customer to automate billing processes and integrate with their existing CRM.

The Cloud Adventures customers are **Cloud Solution Providers** (**CSP**). As the solution is connected to Microsoft Partner Center, they require absolute isolation and a single-tenant deployment. In no way should the API of one customer be able to access the API of another customer.

As cloud solution providers can have multiple hundreds of end customers, the solution needs to be scalable. From a software architecture perspective, the solution is stateless and can be restarted and scaled at any time.

The solution is written in .NET 6 and passed all tests for containerization. From a technical perspective, the developers require several environment variables to be passed:

- >>Billing API address
- >>Billing API key
- >>Azure SQL connection string

As security is high on the list, they also require network traffic to be isolated within a virtual network.

Cloud Adventures currently is configured to run on Azure Container Apps. However, customers are demanding network traffic to remain within the virtual networks managed by Cloud Adventures and also require absolute separation from other customers. Additionally, they require a solution that provides a web application firewall.

That is still quite the use case, and we already addressed all individual requirements in the previous chapter but let's not throw them away. In fact, each time you are looking at new requirements, you always must look at the complete picture. Imagine choosing a different technology based on a single addition. Chances are that the new technology might not even be able to fulfill other requirements.

Please note that the only new addition here will be regarding network traffic. We will add that addition under *security*.

The following are a few bullet point lists that break down the scenario:

- **Security**: From a security perspective, we know a couple of things:
    - Isolation between customers is a hard requirement; the solution must be deployed as *single tenant*. This means deploying a combination of containers for each customer.
    - Cloud Adventures requires integration with virtual networks to keep traffic isolated.
    - Traffic may not leave Cloud Adventures-managed virtual networks.
    - A web application firewall is required.
- **Management**: From a management perspective, we have some additional facts:
    - The ops team is still learning on the job, so complexity must be kept to a minimum.
    - The current focus is on launching customers on the platform. Other than knowledge, time and priority are a constraint for spending time on managing the infrastructure.
    - The solution must be scalable.
- **Financial**: We really only have one requirement here, which is that costs need to be kept to a minimum.
- **Technical**: From a merely technical perspective, we have some other requirements to deal with:
    - There is a frontend that requires access. We will assume this happens over HTTPS.
    - The billing API requires direct access for customers to connect to and integrate with their CRM.
    - Environment variables must be passed upon runtime.

Choosing the right technology

We need to revisit our decisions. Initially, we decided on Azure Container Apps, but the following two requirements have been added to the requirements:

- Traffic may not leave Cloud Adventures-managed virtual networks.
- A web application firewall is required.

Let's see whether that impacts our current decision:

Technology	Yes/no	Motivation
Azure Functions for containers	No	We want to host a frontend and an API per customer. Azure Functions have a completely different use case. However, if it was just for the API, this could be an option.
Azure App Services for Containers	No	Running multiple containers that are scalable requires different containers on different app services or would require Docker Compose, which makes the containers tightly coupled from a scaling perspective.
Azure Container Instances (ACI)	No	Even though ACI provides absolute isolation, it is fairly limited when it comes to scaling. Serving hundreds of customers would potentially hit the limits of the container groups.
Azure Container Apps	No	Even though this was the right choice before, we now need to add a web application firewall as well as network integration. Currently, Azure Container Apps does not provide support for user-defined routes on a networking level that allows us to route traffic to and from a web application firewall.
Azure Kubernetes Services (AKS)	Yes	Technically, yes, and the new requirements justify the management overhead. However, the required knowledge might be too much for the Ops team – not because it's less work now but because we simply do not have another choice. Our challenge now is to keep it simple but be successful!

Table 10.1 – Technology fit table

AKS it is! Now, this is when people can get a little bit nervous, but we will try our best to keep it simple because that is what you need to do with Kubernetes. Keep it simple, start small, and work your way up. Everything we have discussed in the previous chapters of this book can be done on AKS. The question usually is, "*What is the best way to do it?*" Let's show you by starting with visualization.

Visualizing a design

It is always good to visualize a design, as it helps the people who manage it, the developers, and the customer if they require documentation. Let's visualize these requirements at a high level.

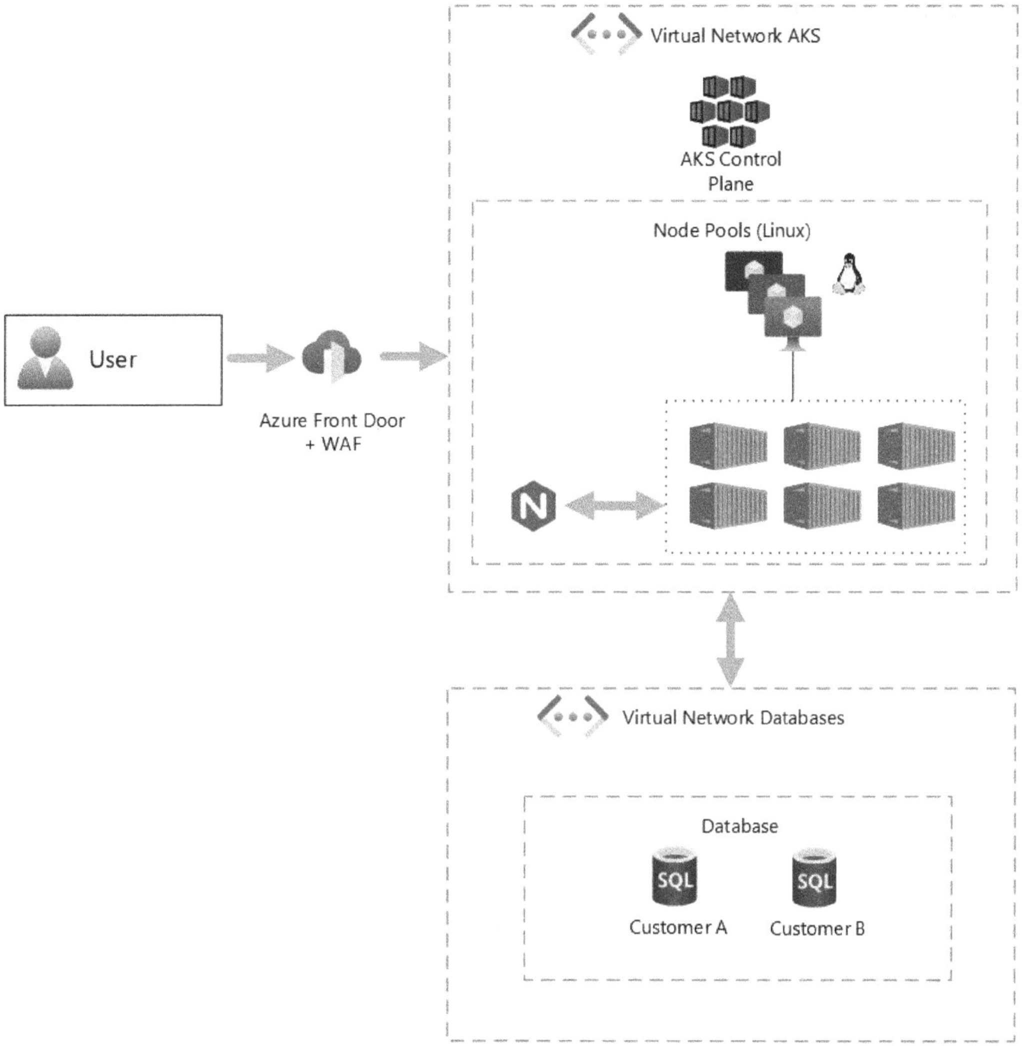
Virtual Network AKS
AKS Control Plane
Node Pools (Linux)
User
Azure Front Door + WAF
Virtual Network Databases
Database
SQL
SQL
Customer A
Customer B

Figure 10.1 – AKS design

What we are showing in this design is the use of separate networks for our AKS infrastructure and databases. Additionally, we will *peer* these networks to make sure these addresses are available over internal addresses and make sure all traffic remains within the Azure Virtual Networks. Additionally, we are using Azure Front Door with a web application firewall configuration to ensure we fulfill that specific requirement. The big question is how we are going to deploy this design. We have the initial infrastructure and then we have the customer-specific infrastructure to deploy when we add future customers. Let's take a look at that in the next section.

Deploying and implementing the solution

It's time to deploy the solution. What we need is the following:

- AKS
- Two containers (the frontend and billing API)
- Isolation per customer (single tenancy)
- Environment variables (the billing API address, billing API key, and Azure SQL connection string)
- Virtual network integration with databases
- Front Door with a web application firewall

Let's start off by creating the needed resources for the AKS cluster and the cluster itself.

Creating the AKS cluster

As always, we need to start off with a resource group. Let's create one using the following command:

```
az group create `
  --name rg-aks-smb `
  --location westeurope
```

Once this command has been completed, you should see some output similar to *Figure 10.2*.

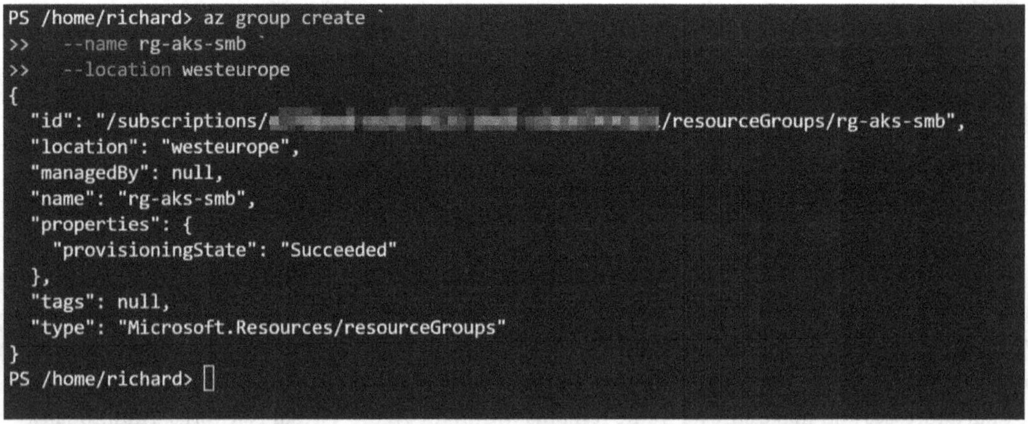
PS /home/richard> az group create `
>> --name rg-aks-smb `
>> --location westeurope
{
"id": "/subscriptions/ /resourceGroups/rg-aks-smb",
"location": "westeurope",
"managedBy": null,
"name": "rg-aks-smb",
"properties": {
"provisioningState": "Succeeded"
},
"tags": null,
"type": "Microsoft.Resources/resourceGroups"
}
PS /home/richard>

Figure 10.2 – Resource group creation completed

With the resource group created, we can now start with the actual resource deployment. As we have a requirement for network communication from the containers to the databases going over the Azure backbone and not the internet, we need to create an Azure **Virtual Network** (**vNet**). To do this, use the following command:

```
az network vnet create `
    --name vnet-aks-smb `
    --resource-group rg-aks-smb `
    --address-prefix 10.240.0.0/24 `
    --subnet-name akscluster `
    --subnet-prefix 10.240.0.0/24 `
    --location westeurope
```

After a few moments, you should have some output like that shown in *Figure 10.3*.

"ipConfigurationProfiles": null,
"ipConfigurations": null,
"name": "akscluster",
"natGateway": null,
"networkSecurityGroup": null,
"privateEndpointNetworkPolicies": "Disabled",
"privateEndpoints": null,
"privateLinkServiceNetworkPolicies": "Enabled",
"provisioningState": "Succeeded",
"purpose": null,
"resourceGroup": "rg-aks-smb",
"resourceNavigationLinks": null,
"routeTable": null,
"serviceAssociationLinks": null,
"serviceEndpointPolicies": null,
"serviceEndpoints": null,
"type": "Microsoft.Network/virtualNetworks/subnets"
}
],
"tags": {},
"type": "Microsoft.Network/virtualNetworks",
"virtualNetworkPeerings": []
}
}

Figure 10.3 – vNet created

It is now time to deploy the fun part – the AKS cluster.

If you have been following along in this book, you may notice that we have added a new parameter to the `aks create` command. This new parameter is called `--vnet-subnet-id`. As the name suggests, it needs the ID of a subnet you want to deploy AKS to. For this example, we will use the `az network vnet subnet show` command to retrieve this ID and store it in the `$AKSsubnetId` variable, which is passed into the `aks create` command.

```
$AKSsubnetId  = az network vnet subnet show --resource-group
rg-aks-smb --vnet-name vnet-aks-smb --name akscluster --query
id


az aks create `
--location westeurope `
--resource-group rg-aks-smb `
--name aks-smb `
--vnet-subnet-id $AKSsubnetId `
--generate-ssh-keys
```

This command can take a few minutes to complete. After the command run, you will see some output similar to *Figure 10.4*.

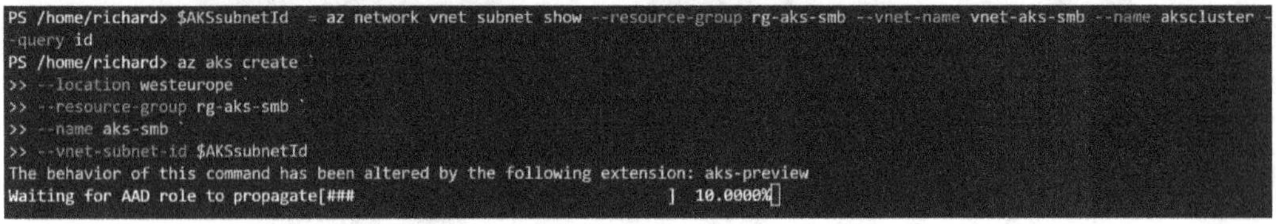
PS /home/richard> $AKSsubnetId = az network vnet subnet show --resource-group rg-aks-smb --vnet-name vnet-aks-smb --name akscluster -
-query id
PS /home/richard> az aks create `
>> --location westeurope `
>> --resource-group rg-aks-smb `
>> --name aks-smb `
>> --vnet-subnet-id $AKSsubnetId
The behavior of this command has been altered by the following extension: aks-preview
Waiting for AAD role to propagate[###] 10.0000%

Figure 10.4 – AKS cluster being created

Once the cluster has been created, you need to connect to it before you can deploy any resources to it. This is super-easy to do with the following command:

```
az aks get-credentials --resource-group rg-aks-smb --name
aks-smb
```

This command will download and merge your cluster login details (context) with the `.kube/config` file in your profile.

PS /home/richard> az aks get-credentials --resource-group rg-aks-smb --name aks-smb
The behavior of this command has been altered by the following extension: aks-preview
Merged "aks-smb" as current context in /home/richard/.kube/config
PS /home/richard>

Figure 10.5 – Getting the AKS cluster credentials

Isolating customers

With the AKS cluster created, it's time to look back at our requirements and, in particular, the network security one. To meet this requirement, we are going to have to peer the AKS vNet to the vNet linked to the customer's database.

> **Important note**
>
> The creation of the database and the vNet it needs is out of the scope of this book, but you can use the following link to create the database if you would like to: `https://docs.microsoft.com/azure/azure-sql/database/private-endpoint-overview?view=azuresql`.

Here's how you can peer the two vNets. First, we need to get the vNet resource IDs. To do that, you can use the following command. You will just need to change the `$customervNetId` details to match your database vNet:

```
$aksvNetId = az network vnet show --name vnet-aks-smb
--resource-group rg-aks-smb --query id --output tsv
$customervNetId = az network vnet show --name vnet-customer1
--resource-group rg-customer1 --query id --output tsv
```

With the two variables set, it's time to peer the two networks. For the peering to work, you actually have to run the same command but with different variables for both vNets.

The first command will set up the peering on the AKS vNet side:

```
az network vnet peering create `
    --name "aks-peer-customer1sql" `
    --resource-group rg-aks-smb `
    --vnet-name vnet-aks-smb `
    --remote-vnet $customervNetId `
    --allow-forwarded-traffic `
    --allow-vnet-access `
    --use-remote-gateways
```

The second command will set up the peering on the database side:

```
az network vnet peering create `
    --name " customer1sql-peer-aks" `
    --resource-group rg-customer1 `
    --vnet-name vnet-customer1 `
    --remote-vnet $aksvNetId `
```

```
    --allow-forwarded-traffic `
    --allow-vnet-access `
    --allow-gateway-transit
```

With the peering all set up, it's time to deploy some Kubernetes resources. One thing to note if you are creating vNet peering via the Azure portal is that once you add them to one vNet, they are automatically added to the other, saving you from configuring twice.

Setting up Ingress

First up is the Ingress controller. This time, we will be deploying it a bit differently. In *Chapter 8*, we deployed the Ingress controller with a public-facing IP address. For this scenario to meet the security requirements, we are going to use Azure Front Door and **Private Link Service** (**PLS**), which means we can deploy an internal Ingress controller that will use an internal Azure load balancer and present the Ingress controller on the vNet.

The following commands will look similar to those from *Chapter 8*, but you may notice some differences. Let's look at them now.

First up, we are using a different namespace. By using the word `internal` in the namespace, we know straight away that this Ingress controller is designed to be an internal one.

The next difference comes with the set commands. They are telling Azure to create an internal load balancer and also to create the PLS for us. This means we do not need to manually create these resources beforehand, as they are managed by AKS, so if we delete the cluster, they will get deleted also.

The last important set command is the `ingress` class. This is useful when using Ingress controllers, especially if you have several on a cluster. By setting an `ingress` class, you are basically telling the Ingress controller to only act on Ingress objects – think manifests here – that match the `ingress` class.

First, we set the namespace variable to `internal-ingress`:

```
$NAMESPACE="internal-ingress"
```

Next, we need to add the `helm` repository to our local machine and update it to ensure we have the latest Helm charts available to us. It's always good to do an update even if you have just added a `helm` repository, as it will update existing ones too:

```
helm repo add ingress-nginx https://kubernetes.github.io/
ingress-nginx
helm repo update
```

Now, it's time to do the Helm installation using the following command. You will notice a lot of `--set` parameters. This overrides the default values of the Helm chart with the ones we need. They could be stored in a Helm values file also, but to make it easier for the demo, we have chosen to use the `--set` parameters:

```
helm install internal-ingress ingress-nginx/ingress-nginx `
--create-namespace `
--namespace $NAMESPACE `
--set controller.service.annotations."service\.beta\.
kubernetes\.io/azure-load-balancer-health-probe-request-path"=/
healthz `
--set controller.service.annotations."service\.beta\.
kubernetes\.io/azure-load-balancer-internal"="true" `
--set controller.service.annotations."service\.beta\.
kubernetes\.io/azure-pls-create"="true" `
--set controller.service.annotations."service\.beta\.
kubernetes\.io/azure-pls-ip-configuration-ip-address-count"="1"
`
--set controller.service.annotations."service\.beta\.
kubernetes\.io/azure-pls-name"="pls-aks" `
--set controller.service.annotations."service\.beta\.
kubernetes\.io/azure-pls-proxy-protocol"="false" `
--set controller.service.annotations."service\.beta\.
kubernetes\.io/azure-pls-visibility"='*' `
--set controller.ingressClass=internal-ingress `
--set controller.ingressClassResource.name=internal-ingress `
--set controller.ingressClassResource.controllerValue="k8s.io/
internal-ingress" `
--set controller.ingressClassResource.enabled=true `
--set controller.ingressClassByName=true `
--set controller.service.externalTrafficPolicy=Local
```

After a short while, the Ingress controller will be deployed, and you should see some output similar to *Figure 10.6*.

```
            backend:
              service:
                name: exampleService
                port:
                  number: 80
            path: /
    # This section is only required if TLS is to be enabled for the Ingress
    tls:
      - hosts:
        - www.example.com
        secretName: example-tls

If TLS is enabled for the Ingress, a Secret containing the certificate and key must also be provided:

  apiVersion: v1
  kind: Secret
  metadata:
    name: example-tls
    namespace: foo
  data:
    tls.crt: <base64 encoded cert>
    tls.key: <base64 encoded key>
  type: kubernetes.io/tls
PS /home/richard>
```

Figure 10.6 – The Ingress controller created

It's always good to make sure the Ingress controller Pods are running and that the service has got an IP address on the subnet. To do this, you can use the following command:

```
kubectl get pods,svc -n internal-ingress
```

In the following output, you will see under Pods that `STATUS` is `Running` and the Ingress controller service has an `EXTERNAL-IP` address on the subnet – in this case, `10.240.0.7`.

```
PS /home/richard> kubectl get pods,svc -n internal-ingress
NAME                                                          READY   STATUS    RESTARTS   AGE
pod/internal-ingress-ingress-nginx-controller-dbc98c796-mc9px   1/1     Running   0          92s

NAME                                                    TYPE           CLUSTER-IP     EXTERNAL-IP   PORT(S)
 AGE
service/internal-ingress-ingress-nginx-controller             LoadBalancer   10.0.242.59    10.240.0.7    80:30590/TCP,443:30278/TCP
 92s
service/internal-ingress-ingress-nginx-controller-admission   ClusterIP      10.0.200.248   <none>        443/TCP
 92s
PS /home/richard>
```

Figure 10.7 – kubectl output

The Ingress controller is now created and the networking is all set up and ready.

Deploying the application

It's time to deploy the applications to the AKS cluster. Let's start off with the API first:

```
kubectl apply -f https://raw.githubusercontent.com/
PacktPublishing/-Up-and-Running-with-Azure-Containers/main/
CH10/api.yaml
```

This will deploy the API deployment, create the service, and also create the Ingress rules for the API.

PS /home/richard> kubectl apply -f https://raw.githubusercontent.com/PacktPublishing/-Up-and-Running-with-Azure-Containers/main/CH10/a
pi.yaml
deployment.apps/billingstatements created
service/billingstatements created
ingress.networking.k8s.io/billingstatements created
PS /home/richard>

Figure 10.8 – API deployment completed

Now, it's time to deploy the frontend:

```
kubectl apply -f https://raw.githubusercontent.com/
PacktPublishing/-Up-and-Running-with-Azure-Containers/main/
CH10/frontend.yaml
```

As with the API, this command will deploy the deployment, create the service, and the Ingress rules, but for the frontend.

PS /home/richard> kubectl apply -f https://raw.githubusercontent.com/PacktPublishing/-Up-and-Running-with-Azure-Containers/main/CH10/f
rontend.yaml
deployment.apps/frontend created
service/frontend created
ingress.networking.k8s.io/frontend created
PS /home/richard>

Figure 10.9 – Frontend deployment completed

If you look at the files before you deploy the frontend, you will see it is basically the same as the one in *Chapter 8*, including the environment variables, but it also has the aforementioned additional `ingress` class.

Let's use `kubectl` to check that everything is up and running:

```
kubectl get deployments,pods,service,ingress
```

You will see that we now have the `frontend` and the `billingstatements` Pods running, the services created, and the Ingress rules pointing to the same internal IP address of the Ingress controller.

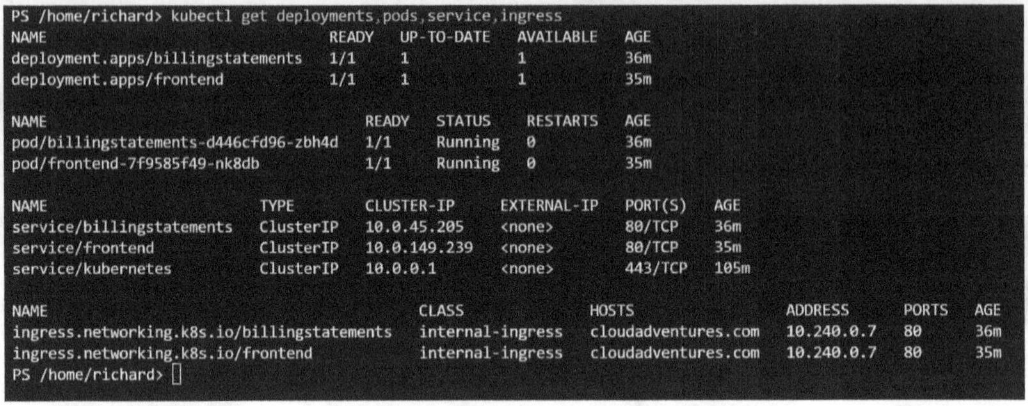
PS /home/richard> kubectl get deployments,pods,service,ingress
NAME READY UP-TO-DATE AVAILABLE AGE
deployment.apps/billingstatements 1/1 1 1 36m
deployment.apps/frontend 1/1 1 1 35m
NAME READY STATUS RESTARTS AGE
pod/billingstatements-d446cfd96-zbh4d 1/1 Running 0 36m
pod/frontend-7f9585f49-nk8db 1/1 Running 0 35m
NAME TYPE CLUSTER-IP EXTERNAL-IP PORT(S) AGE
service/billingstatements ClusterIP 10.0.45.205 <none> 80/TCP 36m
service/frontend ClusterIP 10.0.149.239 <none> 80/TCP 35m
service/kubernetes ClusterIP 10.0.0.1 <none> 443/TCP 105m
NAME CLASS HOSTS ADDRESS PORTS AGE
ingress.networking.k8s.io/billingstatements internal-ingress cloudadventures.com 10.240.0.7 80 36m
ingress.networking.k8s.io/frontend internal-ingress cloudadventures.com 10.240.0.7 80 35m
PS /home/richard>

Figure 10.10 – kubectl output

Everything is ready to test, except we only have an internal IP address. The sites are not accessible to the internet. This is where `AzureFrontDoor` comes in!

Configuring Azure Front Door

To access the websites from the internet and to match the requirement of using a **Web Application Firewall** (**WAF**), we need to deploy a Front Door. To do that, use the following commands:

```
az afd profile create `
--profile-name fd-smb `
--resource-group rg-aks-smb `
--sku Premium_AzureFrontDoor
```

After a short while, the `AzureFrontDoor` resource will be created, but there is now configuration to do.

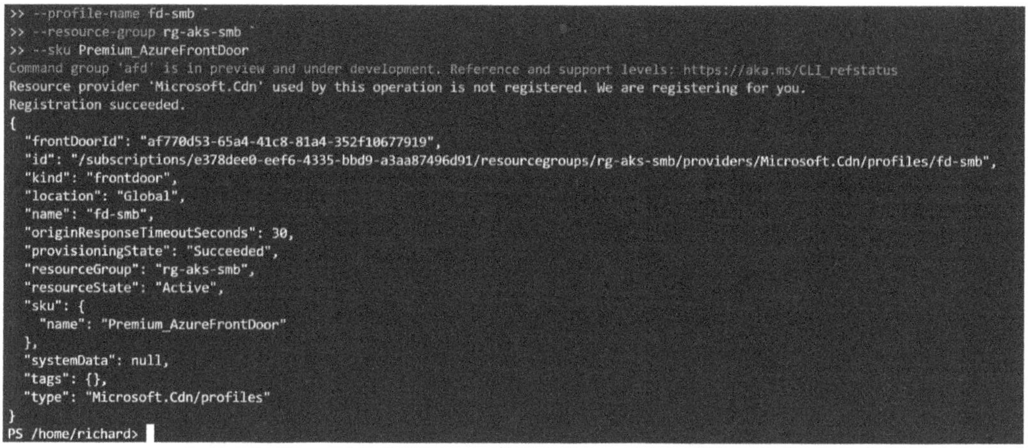
>> --profile-name fd-smb `
>> --resource-group rg-aks-smb `
>> --sku Premium_AzureFrontDoor
Command group 'afd' is in preview and under development. Reference and support levels: https://aka.ms/CLI_refstatus
Resource provider 'Microsoft.Cdn' used by this operation is not registered. We are registering for you.
Registration succeeded.
{
"frontDoorId": "af770d53-65a4-41c8-81a4-352f10677919",
"id": "/subscriptions/e378dee0-eef6-4335-bbd9-a3aa87496d91/resourcegroups/rg-aks-smb/providers/Microsoft.Cdn/profiles/fd-smb",
"kind": "frontdoor",
"location": "Global",
"name": "fd-smb",
"originResponseTimeoutSeconds": 30,
"provisioningState": "Succeeded",
"resourceGroup": "rg-aks-smb",
"resourceState": "Active",
"sku": {
"name": "Premium_AzureFrontDoor"
},
"systemData": null,
"tags": {},
"type": "Microsoft.Cdn/profiles"
}
PS /home/richard>

Figure 10.11 – AzureFrontDoor created

> **Important note**
>
> If this is the first time you have deployed `AzureFrontDoor`, you will need to register the resource provider, which can take some time.

Now comes the Front Door configuration. First up is the endpoint. To create this, use the following command:

```
az afd endpoint create `
--profile-name fd-smb `
--resource-group rg-aks-smb `
--endpoint-name cloudadventures `
--enabled-state Enabled
```

Within a few seconds, the endpoint will be created.

PS /home/richard> az afd endpoint create `
>> --profile-name fd-smb `
>> --resource-group rg-aks-smb `
>> --endpoint-name cloudadventures `
>> --enabled-state Enabled
Command group 'afd' is in preview and under development. Reference and support levels: https://aka.ms/CLI_refstatus
{
"autoGeneratedDomainNameLabelScope": "TenantReuse",
"deploymentStatus": "NotStarted",
"enabledState": "Enabled",
"hostName": "cloudadventures-bdhuatepepfzb5cz.z01.azurefd.net",
"id": "/subscriptions/e378dee0-eef6-4335-bbd9-a3aa87496d91/resourcegroups/rg-aks-smb/providers/Microsoft.Cdn/profiles/fd-smb/afdendp
oints/cloudadventures",
"location": "Global",
"name": "cloudadventures",
"profileName": null,
"provisioningState": "Succeeded",
"resourceGroup": "rg-aks-smb",
"systemData": null,
"tags": {},
"type": "Microsoft.Cdn/profiles/afdendpoints"
}
PS /home/richard>

Figure 10.12 – Endpoint created

With the endpoint created, we can move on to the origin group. An origin group is a bit like a folder to hold information about your origin (or, in this case, the AKS cluster):

```
az afd origin-group create `
--resource-group rg-aks-smb `
--profile-name fd-smb `
--origin-group-name fd-origin-smb `
--probe-request-type GET `
--probe-protocol Http `
--probe-interval-in-seconds 60 `
--probe-path / `
--sample-size 4 `
--successful-samples-required 3 `
--additional-latency-in-milliseconds 50
```

Once this command has finished, you should have some output similar to *Figure 10.13*.

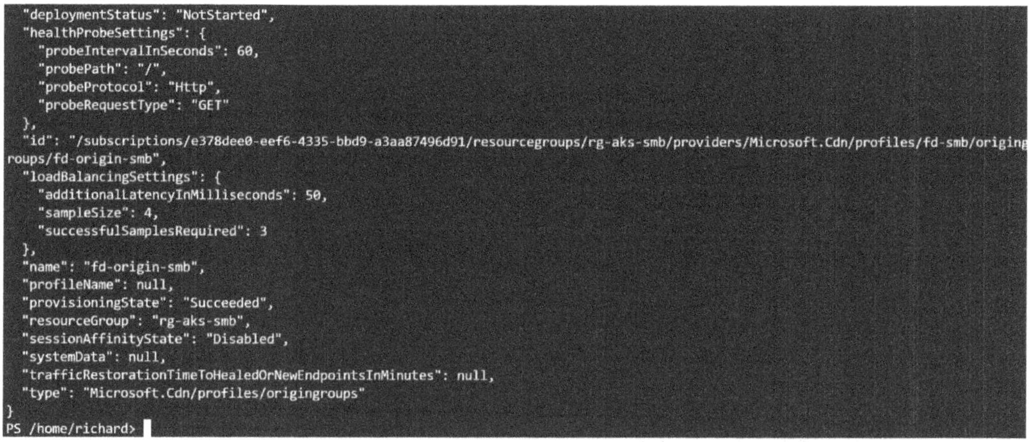
"deploymentStatus": "NotStarted",
"healthProbeSettings": {
"probeIntervalInSeconds": 60,
"probePath": "/",
"probeProtocol": "Http",
"probeRequestType": "GET"
},
"id": "/subscriptions/e378dee0-eef6-4335-bbd9-a3aa87496d91/resourcegroups/rg-aks-smb/providers/Microsoft.Cdn/profiles/fd-smb/originggroups/fd-origin-smb",
"loadBalancingSettings": {
"additionalLatencyInMilliseconds": 50,
"sampleSize": 4,
"successfulSamplesRequired": 3
},
"name": "fd-origin-smb",
"profileName": null,
"provisioningState": "Succeeded",
"resourceGroup": "rg-aks-smb",
"sessionAffinityState": "Disabled",
"systemData": null,
"trafficRestorationTimeToHealedOrNewEndpointsInMinutes": null,
"type": "Microsoft.Cdn/profiles/origingroups"
}
PS /home/richard>

Figure 10.13 – Origin group created

Now it's time to add the origin, or AKS cluster in this case, but first, we need some more information. We need the PLS alias and the `pls` ID both of which were created for us when we deployed the internal Ingress controller. But don't worry – at the top of the following command, we will get both the PLS alias and ID and save them as variables for the `origin create` command:

```
$plsAlias = az network private-link-service show `
--resource-group "mc_rg-aks-smb_aks-smb_westeurope" `
--name "pls-aks" `
--query alias -o tsv
$plsId = az network private-link-service show `
--resource-group "mc_rg-aks-smb_aks-smb_westeurope" `
--name "pls-aks" `
--query id -o tsv
```

Now that we have both IDs, we can use the following command to create the origin:

```
az afd origin create `
--resource-group rg-aks-smb `
--profile-name fd-smb `
--host-name $plsAlias `
--origin-group-name fd-origin-smb `
--origin-name pls-aks `
--origin-host-header cloudadventures.com `
```

```
--priority 1 `
--weight 1000 `
--enabled-state Enabled `
--http-port 80 `
--enable-private-link true `
--private-link-location westeurope `
--private-link-resource $plsId `
--private-link-request-message "please approve"
```

Once this has finished, you will have some output like that shown in *Figure 10.14*.

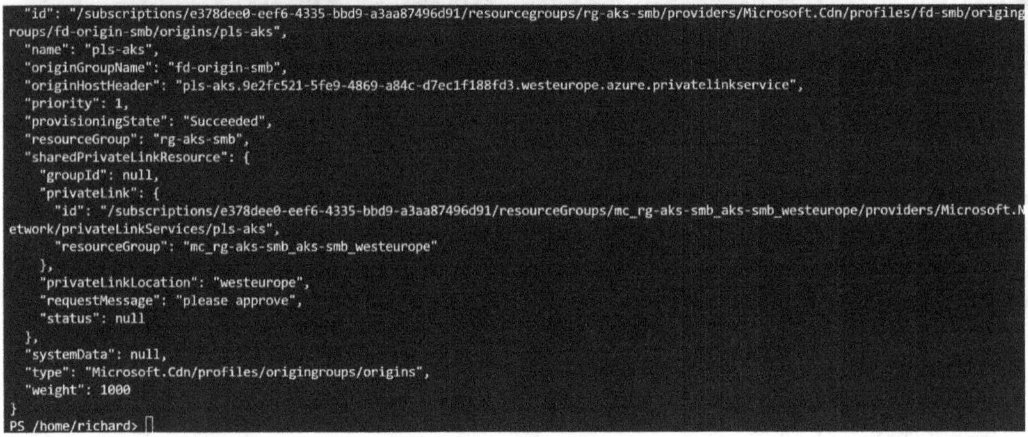
"id": "/subscriptions/e378dee0-eef6-4335-bbd9-a3aa87496d91/resourcegroups/rg-aks-smb/providers/Microsoft.Cdn/profiles/fd-smb/originG
roups/fd-origin-smb/origins/pls-aks",
"name": "pls-aks",
"originGroupName": "fd-origin-smb",
"originHostHeader": "pls-aks.9e2fc521-5fe9-4869-a84c-d7ec1f188fd3.westeurope.azure.privatelinkservice",
"priority": 1,
"provisioningState": "Succeeded",
"resourceGroup": "rg-aks-smb",
"sharedPrivateLinkResource": {
"groupId": null,
"privateLink": {
"id": "/subscriptions/e378dee0-eef6-4335-bbd9-a3aa87496d91/resourceGroups/mc_rg-aks-smb_aks-smb_westeurope/providers/Microsoft.N
etwork/privateLinkServices/pls-aks",
"resourceGroup": "mc_rg-aks-smb_aks-smb_westeurope"
},
"privateLinkLocation": "westeurope",
"requestMessage": "please approve",
"status": null
},
"systemData": null,
"type": "Microsoft.Cdn/profiles/origingroups/origins",
"weight": 1000
}
PS /home/richard>

Figure 10.14 – Origin created

You may notice in the command that we have passed in a private link message. This is needed, as we cannot auto-approve the private link request. In fact, we have to go to the Azure portal to approve the connection from Front Door:

1. In the Azure portal (`https://portal.azure.com`), search for your private link name and click it.

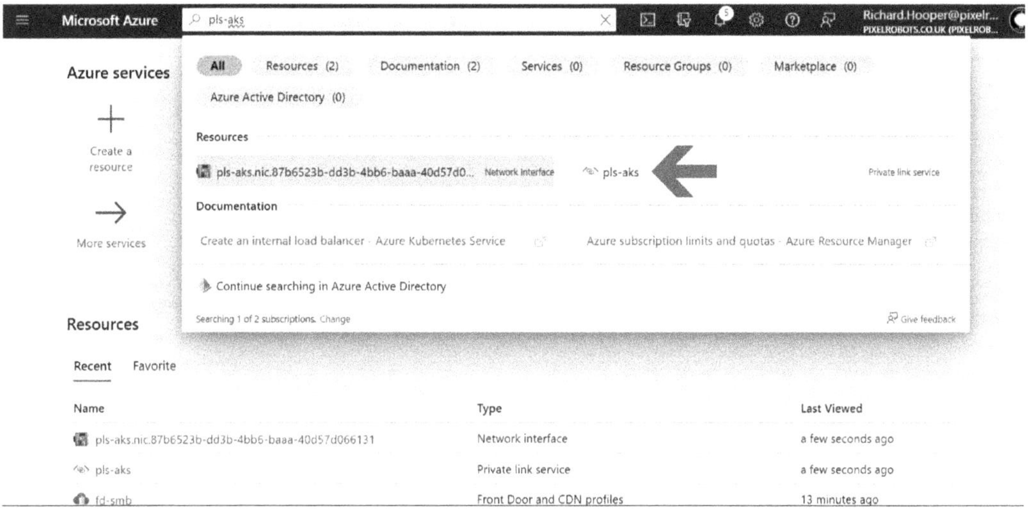
Microsoft Azure
pls-aks
Richard.Hooper@pixelr...
PIXELROBOTS.CO.UK (PIXELROB...
Azure services
Create a resource
More services
All
Resources (2)
Documentation (2)
Services (0)
Resource Groups (0)
Marketplace (0)
Azure Active Directory (0)
Resources
pls-aks.nic.87b6523b-dd3b-4bb6-baaa-40d57d0...
Network interface
pls-aks
Private link service
Documentation
Create an internal load balancer - Azure Kubernetes Service
Azure subscription limits and quotas - Azure Resource Manager
Continue searching in Azure Active Directory
Searching 1 of 2 subscriptions. Change
Give feedback
Resources
Recent
Favorite
Name
Type
Last Viewed
pls-aks.nic.87b6523b-dd3b-4bb6-baaa-40d57d066131
Network interface
a few seconds ago
pls-aks
Private link service
a few seconds ago
fd-smb
Front Door and CDN profiles
13 minutes ago

Figure 10.15 – Searching for the private link in the Azure portal

2. In the private link blade, click on **Private endpoint connections**.

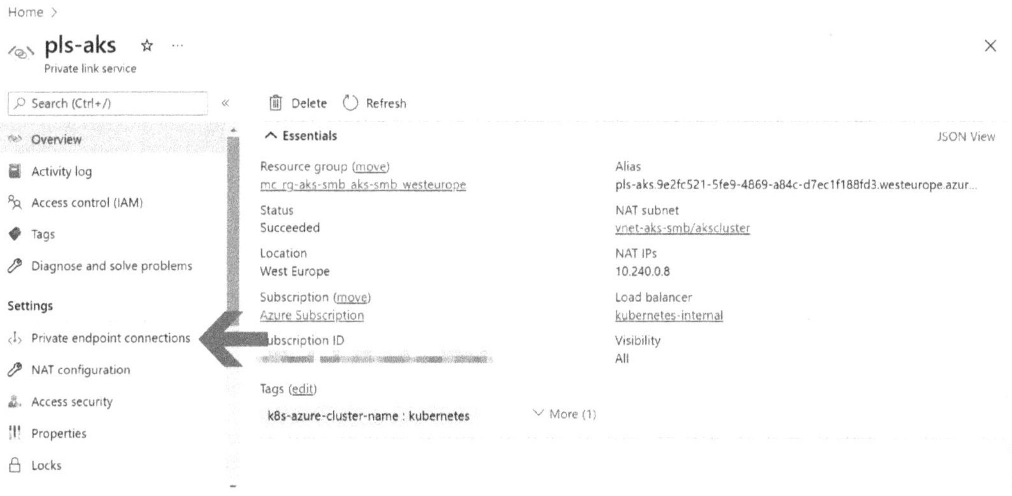
Home >
pls-aks
Private link service
Search (Ctrl+/)
Delete
Refresh
Overview
Activity log
Access control (IAM)
Tags
Diagnose and solve problems
Settings
Private endpoint connections
NAT configuration
Access security
Properties
Locks
Essentials
JSON View
Resource group (move)
mc_rg-aks-smb_aks-smb_westeurope
Status
Succeeded
Location
West Europe
Subscription (move)
Azure Subscription
Subscription ID
Tags (edit)
k8s-azure-cluster-name : kubernetes
More (1)
Alias
pls-aks.9e2fc521-5fe9-4869-a84c-d7ec1f188fd3.westeurope.azur...
NAT subnet
vnet-aks-smb/akscluster
NAT IPs
10.240.0.8
Load balancer
kubernetes-internal
Visibility
All

Figure 10.16 – Clicking Private endpoint connections

3. Check the connection name box and then click **Approve** | **Yes**. After a short while, this will be approved.

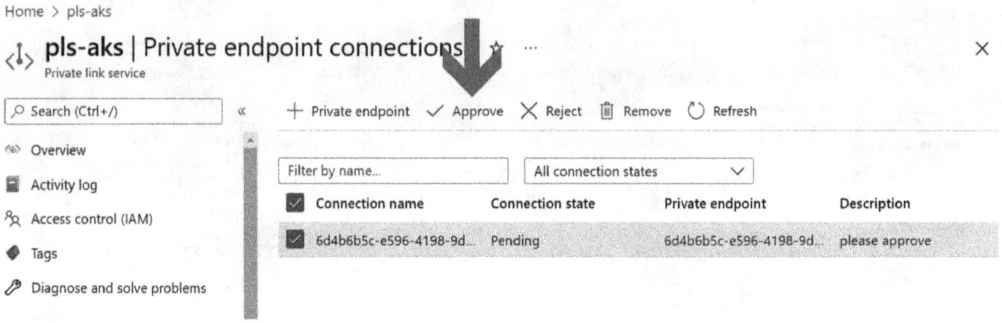
Home > pls-aks
pls-aks | Private endpoint connections
Private link service
Search (Ctrl+/)
Overview
Activity log
Access control (IAM)
Tags
Diagnose and solve problems
+ Private endpoint
Approve
Reject
Remove
Refresh
Filter by name...
All connection states
Connection name
Connection state
Private endpoint
Description
6d4b6b5c-e596-4198-9d...
Pending
6d4b6b5c-e596-4198-9d...
please approve

Figure 10.17 – Clicking Approve

It's now time to create a route from Front Door to our origin (the AKS cluster). To do this, use the following command:

```
az afd route create `
--resource-group rg-aks-smb `
--profile-name fd-smb `
--endpoint-name cloudadventures `
--forwarding-protocol MatchRequest `
--route-name aks-route `
--https-redirect Disabled `
--origin-group fd-origin-smb `
--supported-protocols Http `
--link-to-default-domain Enabled
```

Once completed, you should have some JSON output, as shown in *Figure 10.18*.

"httpsRedirect": "Disabled",
"id": "/subscriptions/e378dee0-eef6-4335-bbd9-a3aa87496d91/resourcegroups/rg-aks-smb/providers/Microsoft.Cdn/profiles/fd-smb/afdendpoints/cloudadventures/routes/aks-route",
"linkToDefaultDomain": "Enabled",
"name": "aks-route",
"originGroup": {
"id": "/subscriptions/e378dee0-eef6-4335-bbd9-a3aa87496d91/resourceGroups/rg-aks-smb/providers/Microsoft.Cdn/profiles/fd-smb/originGroups/fd-origin-smb",
"resourceGroup": "rg-aks-smb"
},
"originPath": null,
"patternsToMatch": [
"/*"
],
"provisioningState": "Succeeded",
"resourceGroup": "rg-aks-smb",
"ruleSets": [],
"supportedProtocols": [
"Http"
],
"systemData": null,
"type": "Microsoft.Cdn/profiles/afdendpoints/routes"
}
PS /home/richard>

Figure 10.18 – Route created

That's the Front Door created and configured, but we still have not met the requirement of needing a WAF. But don't worry – that's easy to do with the following command:

```
az network front-door waf-policy create `
--name wafsmb `
--resource-group rg-aks-smb `
--sku Premium_AzureFrontDoor `
--disabled false `
--mode Prevention
```

You may have noticed that we have enabled prevention mode. This is okay for this example, but we would recommend using detection mode first for production, and then looking at the logs and making any changes to ensure your application works before you enable prevention mode.

"managedRules": {
"managedRuleSets": []
},
"name": "wafsmb",
"policySettings": {
"customBlockResponseBody": null,
"customBlockResponseStatusCode": null,
"enabledState": "Enabled",
"mode": "Prevention",
"redirectUrl": null,
"requestBodyCheck": "Enabled"
},
"provisioningState": "Succeeded",
"resourceGroup": "rg-aks-smb",
"resourceState": "Enabled",
"routingRuleLinks": [],
"securityPolicyLinks": [],
"sku": {
"name": "Premium_AzureFrontDoor"
},
"tags": {},
"type": "Microsoft.Network/frontdoorwebapplicationfirewallpolicies"
}
PS /home/richard>

Figure 10.19 – The WAF policy created

With the policy created, it is now time to assign it to the endpoint we created earlier in this chapter. You will need to change the following command to your own subscription ID:

```
az afd security-policy create `
--resource-group rg-aks-smb `
--profile-name fd-smb `
--security-policy-name wafsmb `
--domains "/subscriptions/<subscriptionId>/resourcegroups/
rg-aks-smb/providers/Microsoft.Cdn/profiles/fd-smb/
afdEndpoints/cloudadventures" `
--waf-policy "/subscriptions/<subscriptionId>/
resourceGroups/rg-aks-smb/providers/Microsoft.Network/
frontdoorwebapplicationfirewallpolicies/wafsmb"
```

If everything goes according to plan, you should have some output like *Figure 10.20*.

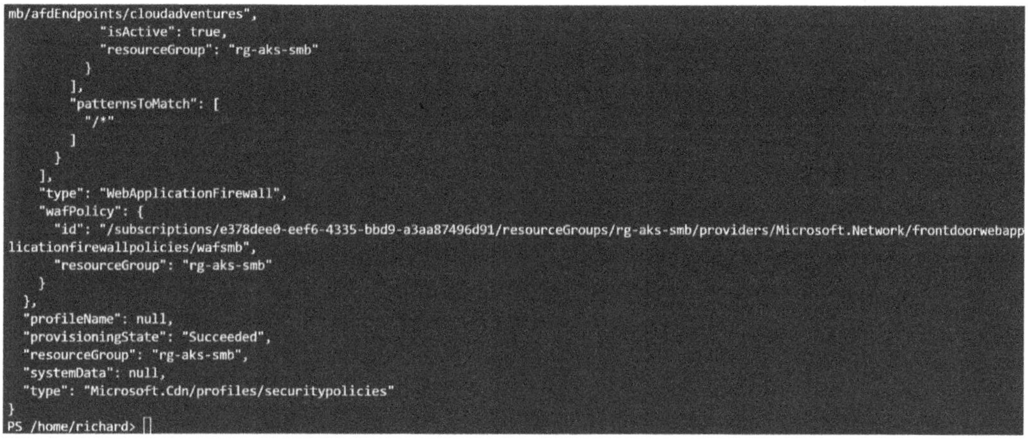

mb/afdEndpoints/cloudadventures",
"isActive": true,
"resourceGroup": "rg-aks-smb"
}
],
"patternsToMatch": [
"/*"
]
}
],
"type": "WebApplicationFirewall",
"wafPolicy": {
"id": "/subscriptions/e378dee0-eef6-4335-bbd9-a3aa87496d91/resourceGroups/rg-aks-smb/providers/Microsoft.Network/frontdoorwebapp
licationfirewallpolicies/wafsmb",
"resourceGroup": "rg-aks-smb"
}
},
"profileName": null,
"provisioningState": "Succeeded",
"resourceGroup": "rg-aks-smb",
"systemData": null,
"type": "Microsoft.Cdn/profiles/securitypolicies"
}
PS /home/richard>

Figure 10.20 – The WAF policy assigned to the endpoint

Azure Front Door is now configured!

Testing the websites

With all that configuration, it's finally time to check whether the sites are working. But what's the URL? We can get that by using the following:

```
az afd endpoint show --profile-name fd-smb --endpoint-name
cloudadventures --resource-group rg-aks-smb --query hostName -o
tsv
```

Instantly, you should see a web address ending in `azurefd.net`.

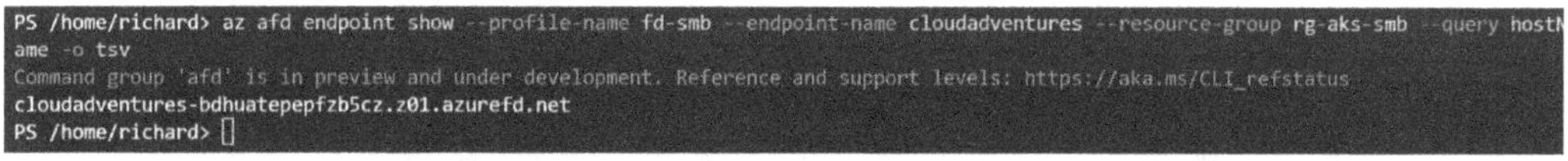

PS /home/richard> az afd endpoint show --profile-name fd-smb --endpoint-name cloudadventures --resource-group rg-aks-smb --query hostN
ame -o tsv
Command group 'afd' is in preview and under development. Reference and support levels: https://aka.ms/CLI_refstatus
cloudadventures-bdhuatepepfzb5cz.z01.azurefd.net
PS /home/richard>

Figure 10.21 – The website address from Azure Front Door

Copy the address to your clipboard and paste it into your web browser. You should see the Cloud Adventures website.

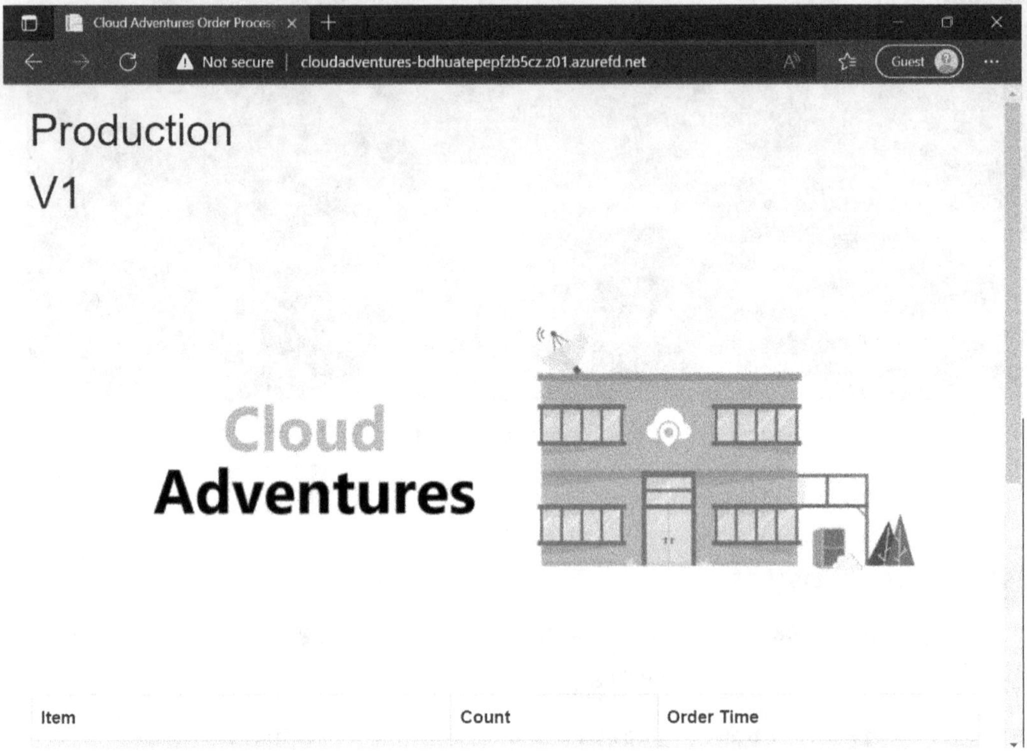
Cloud Adventures Order Process
Not secure
cloudadventures-bdhuatepepfzb5cz.z01.azurefd.net
Guest
Production
V1
Cloud
Adventures
Item
Count
Order Time

Figure 10.22 – The Cloud Adventures website

Let's check that the `billingstatements` API is still reachable. Just add `/billingstatement` to the end of the URL. If all works correctly, you should see a web page showing some JSON.

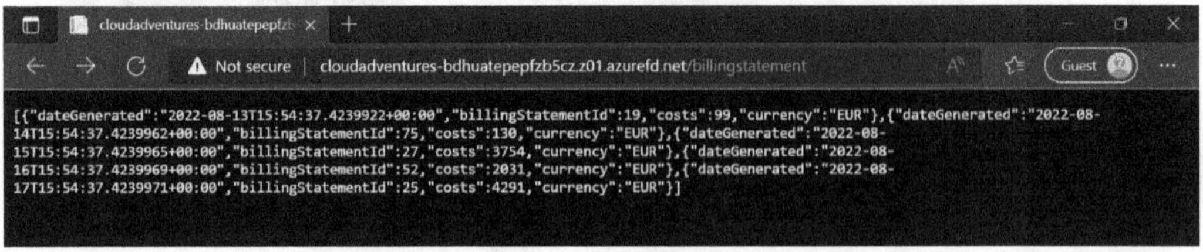
cloudadventures-bdhuatepepfzb
Not secure
cloudadventures-bdhuatepepfzb5cz.z01.azurefd.net/billingstatement
Guest
[{"dateGenerated":"2022-08-13T15:54:37.4239922+00:00","billingStatementId":19,"costs":99,"currency":"EUR"},{"dateGenerated":"2022-08-14T15:54:37.4239962+00:00","billingStatementId":75,"costs":130,"currency":"EUR"},{"dateGenerated":"2022-08-15T15:54:37.4239965+00:00","billingStatementId":27,"costs":3754,"currency":"EUR"},{"dateGenerated":"2022-08-16T15:54:37.4239969+00:00","billingStatementId":52,"costs":2031,"currency":"EUR"},{"dateGenerated":"2022-08-17T15:54:37.4239971+00:00","billingStatementId":25,"costs":4291,"currency":"EUR"}]

Figure 10.23 – The billing statement API website

> **Important note**
>
> In real life, you would use your own certificate or take advantage of tools such as Let's Encrypt and cert-manager. This is out of the scope of this book, but if you would like to read more about cert-manager, then please look at `https://cert-manager.io/docs/`.

Learnings and future decisions

What we have learned is that customers demand change. Especially when you are growing as a company, your customers will change as well. You will attract bigger customers that have different requirements. And that is perfectly fine. It just means you need to adjust a little.

The biggest thing to learn here, however, is that out-of-the-box solutions such as Azure Container Apps (or ACI) provide great features when you are starting out. However, when requirements change and you need to adjust, these services are not as adjustable as you would like. AKS does come with more complexity – we will be the first people to admit that – but it also comes with great flexibility, and what we have seen is that if you start small and gradually move onward, it doesn't have to be that complicated. Yes, there is a learning curve, but dealing with that comes down to decision-making. If you decide that AKS is going to be your next option early in the process, just gradually move on. Learn by doing and don't implement things you don't need (yet). Remember – small steps will result in big leaps.

Additionally, what we have learned is that when you really want to control what you're doing and are using more complex technologies, YAML is a must-know language. It's not really difficult, but you need to understand it, read it, and be able to debug it.

Summary

In this chapter, we have learned that growth in your business can mean a change in technology. It's really about how you plan for this and even whether you can accept it. A decision you made a year ago can very well need to be revisited – accept that. Look at it like this – the fact that you had to revisit this decision means your company is doing well. There is no shame in that!

We have seen that out-of-the-box technologies are fine until you have to deal with specific requirements. Once that happens, the answer is almost always AKS – not just because we love AKS but because it also provides flexibility. And if we really just go by the requirements, it is not that difficult to use. When it comes to using Kubernetes, it only becomes difficult when you want to go from 0–100. Just use what you need.

In the next chapter, we will look at an enterprise scenario. What about scaling? What about management and monitoring? There is no single answer to those; there are actually many ways to go about them, and we can't wait to share our best practices with you!

11
Container Technologies for Enterprises

You start attracting bigger customers and your focus is shifting to governance and compliance. Not that you haven't thought of that before, but we can no longer get away with "it likely shouldn't cause a problem." Now, that is somewhat of an ambiguous sentence, but it is a result of not knowing your infrastructure is ready for your biggest customers.

Still, it doesn't have to be perfect. You can still improve your solution as you go. In fact, if you don't, this will result in a whole world of problems down the line. But there are some hygiene factors you want to get right before landing those bigger customers. Not because the software solution itself is that different but because those customers generally have a department that focuses on governance and compliance and has elaborate supplier management processes.

We always like to see the bright side of that. It forces us to rethink what we have built and to adjust where necessary.

In the previous chapter, we asked you not to delete what was deployed, as we are going to build on top of that because we don't have to throw everything away and start over to satisfy our customers.

So, what do enterprise customers really want? The short answer is *nothing to be worried about*. AKS has got your back.

In this chapter, we're going to go over the following topics:

- Scenario and design
- Deploying and implementing the solution
- Learnings and future decisions

Cloud Adventures webshop – scenario and design

As in the previous chapters, let's continue and build on top of the scenario that we presented in the previous chapter. We have already repeated the use case several times, and to prevent too much repetition, we will only add the new requirements.

> An e-commerce company called Cloud Adventures provides a webshop for its customers that allows them to buy Office 365 licenses. The solution consists of a frontend portal and a backend API that processes the billing. Additionally, the backend API allows direct access by the customer to automate billing processes and integrate with their existing CRM.
>
> Cloud Adventures has proven that it can run the solution and provide a reliable configuration on Microsoft Azure. Its customer base is growing and, even more importantly, it is adding customers of bigger enterprise sizes and a higher level of organizational maturity.
>
> These customers come with additional requirements; albeit technical, these requirements are slightly different. Enterprise customers require more governance-related configurations, mostly in the realm of monitoring and security. These customers require thorough monitoring and secure handling of access to the Azure platform itself. Additionally, the Cloud Adventures Ops team has challenges keeping up with management tasks and deploying the vast number of new customers, as they have been storing environment variables and secrets directly on the AKS cluster. This requires them to always modify existing YAML and apply it to the cluster. In the past couple of weeks, this has resulted in a human error (modifying the secrets of the wrong customer) but even more importantly, the maintenance load is significant.
>
> These new requirements have been asked for by Cloud Adventures' customers for a little over 2 months now. They have been struggling to implement the requirements, as the current environment is already running in production. However, Cloud Adventures does see these new changes as a high priority as they will help sales for new customers and also improve the governance of the current environment. In any case, they do not want to rebuild the existing cluster and perform migration for existing customers.

The customer requirements in the preceding use case are not uncommon and are a typical result of focusing on launching the platform as soon as possible. Providing the functionality to the customer that directly benefits the solution is the biggest focus for every software company. However, that results in what we can refer to as *governance debt*. Eventually, we need to make sure all hygiene factors, such as monitoring and security, are in place. Believe us when we say that this will not only improve reliability and provide insights, it will also give your Ops team and service level managers peace of mind.

The new requirements

From the use case, we can distill a couple more requirements in addition to the requirements established in the previous chapter.

> **Important note**
>
> As we are building on the previous use case and did not delete the previously built cluster, we are not deep diving into the existing requirements. We are adding to the things we already have.

We are seeing the following additions to our requirements:

- Customers require insights and monitoring.
- Secure access to the cluster is required.
- Secure interaction with the Azure platform is required.
- The current secrets management on the cluster itself is not sufficient and requires a change of tactics.

Luckily, this is nothing we can't solve by making some additions to our cluster. Let's take a look at the technologies we are going to implement.

Azure Monitor/Container insights

Monitoring the behavior of your cluster is key. Even though the basic information on cluster operations can be pulled directly from the Azure portal, it doesn't provide us with enough information to proactively act on potential issues.

Being able to monitor your AKS clusters, nodes, and workloads is very important at any stage of your business. When you were a start-up or small to medium business, you could probably have gotten away with logging into a text file, or even going into the Pods to get the logs directly.

In our scenario, we have a single-tenant application that has two Pods. So, if we had 100 customers, that's 200 Pods all logging into a different text file or not at all. And remember, we want to keep the Ops overhead to a minimum.

By using Container insights, you can store not only the AKS cluster logs in a central place but also your application logs. This makes troubleshooting a lot easier. And as it is an Azure resource, you can use RBAC to allow your support staff to only see the logs from this central point and have no access to the AKS cluster. Why is this important? Having centralized logging and providing the application itself to log to the same environment that the cluster operations are being logged to, we can analyze the behavior of both the application and infrastructure. We can determine how one impacts another and provide valuable information to both the development and the Ops teams.

Managed identity

A big issue in security is dealing with service accounts, service principals, and keys. People have been using service principals in Azure Active Directory for a very long time. These service principals come with a **client ID** and a **client secret** that we can use to authenticate and receive access to the Azure subscription and its resources.

The catch here is the client secret. Upon creation, we choose a lifetime for that secret. We all know, and we all agree, that a shorter lifetime for that secret is best. But 99% of the time, we choose a lifetime that will probably outlast the lifetime of the product (multiple years). The longer a secret exists and remains the same, the bigger the chance is that it will be, or has been, compromised. The absolute best practice we all agree on in IT is key rotation. However, we still choose a very long lifetime for that secret… why? Well, imagine the key expires and we forgot to monitor it; this might bring the entire application to a halt.

Microsoft has introduced a solution for this called **managed identities**. Basically, we are provisioning a *managed* service principal, which we allow the cluster or any other resource to use. We no longer have to worry about the rotation of keys. In fact, we have no clue about the key as it is automatically rotated and passed to the service using the managed identity whenever it needs it.

This is a far better way to deal with secrets than traditional service principals.

Azure AD authentication

When you deploy an AKS cluster, by default, a local admin account is created in the cluster that has the cluster-admin role, the highest privileges you can have in an AKS cluster. It is highly recommended you disable this local admin account and enable Azure AD integration for your cluster. Why? If someone were ever to share their `KubeConfig` file with those credentials, it no longer matters what kind of security countermeasures you implemented. Malicious actors in possession of your `KubeConfig` file will have a field day. Let's also not forget user management. If you run multiple clusters and you have multiple `KubeConfig` files being used throughout your Azure estate, what if an employee leaves the company? How are you going to be sure they no longer have access to the cluster? Not by walking up to their workstation and manually deleting the `KubeConfig` file!

What we want is the same thing we want with every identity provider. We want multiple levels of security for authentication, and we want **Multi-Factor Authentication** (**MFA**). On top of that, we want centralized user management.

This is exactly what Microsoft provides when using AKS. We want to disable the local Kubernetes admin account, and we want this additional authentication layer on top of the `KubeConfig` file.

Key Vault and AKS

Environment variables and secrets are most commonly configured directly in the YAML files or sometimes created by using the `kubectl` commands. Ideally, you separate customers per namespace but that comes with a price when it comes to secrets management. Secrets are scoped to the context of a namespace. That means secrets in every namespace. That also means a lot of management. We can do that better.

On Azure, we traditionally store secrets in Azure Key Vault. It is a good way to centralize the management of secrets. As Azure Key Vault comes with both RBAC and additional authorization specific to Azure Key Vault, this already sounds better than passing secrets as plain text into a secret.

We just discussed managed identities as well. And guess what? We can use the managed identities linked to AKS to access the Key Vault without requiring any human interaction or credentials at all.

You may think it is simply logical to deploy these features to AKS but the reality is different. When we first deploy a solution as a start-up or even as a small or medium business, we focus on providing direct value to our end customers. Features such as Azure AD, managed identities, and Azure Key Vault don't really sound like they directly add value. However, not doing so will open you up to risks such as human error and outside attacks. In the end, that can be a very expensive scenario. Normally, we realize this once we start growing bigger as a company. The risk levels increase.

By configuring these services, we are indirectly adding value for our customers and also to our own business. Let's take a look at how these features fit into our design.

Visualizing a design

We are going to make an addition to our existing design. It might look similar to the visualization from the previous chapter, but we are adding Azure Container insights, Azure Key Vault, Azure AD, and managed identity. You could say that when creating a design, these will always be present.

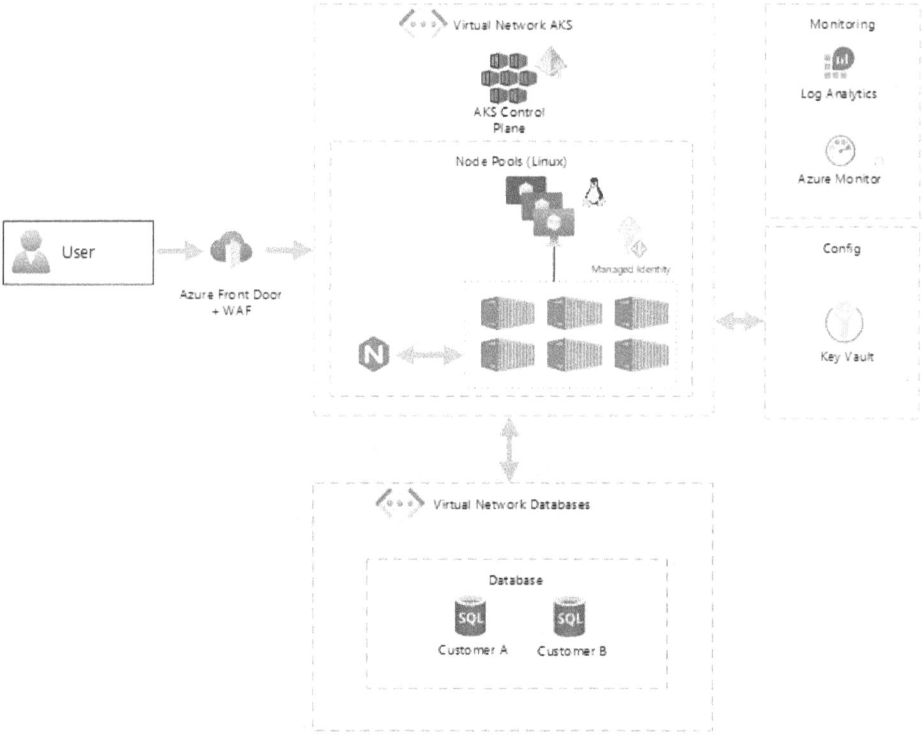
Virtual Network AKS
AKS Control Plane
Node Pools (Linux)
Managed Identity
Monitoring
Log Analytics
Azure Monitor
Config
Key Vault
User
Azure Front Door + WAF
Virtual Network Databases
Database
Customer A
Customer B

Figure 11.1 – AKS design

What we visualized in this design is to use separate networks for our AKS infrastructure and our databases. Additionally, we will *peer* these networks to make sure these addresses are available over internal addresses and make sure all traffic remains within the Azure virtual networks. Additionally, we are using Azure Front Door with a **Web Application Firewall (WAF)** configuration to ensure that we fulfill that specific requirement.

The big question is: How are we going to deploy this design? We have the initial infrastructure and then we have the customer-specific infrastructure to deploy when we add future customers. Let us take a look at that in the next section.

Deploying and implementing the solution

Let's start deploying the additions we specified earlier to our cluster.

Enabling Azure Monitor/Container insights

Before you can enable Container insights, you need a Log Analytics workspace. A Log Analytics workspace is an Azure resource for storing logs. You can store multiple types of logs in a workspace too. For Container insights, you can have not only the Kubernetes logs but also your application's `stdout` and `stderr` logs. Let's look at creating a Log Analytics workspace.

In the real world, you may want to deploy the Log Analytics workspace into its own resource group or one of your choosing. As we already have the resource group for our AKS cluster, we can go ahead and use it for the Log Analytics workspace too:

```
az monitor log-analytics workspace create `
            --resource-group rg-aks-smb   `
            --workspace-name log-aks-smb  `
            --location westeurope
```

This command normally takes around a minute to complete.

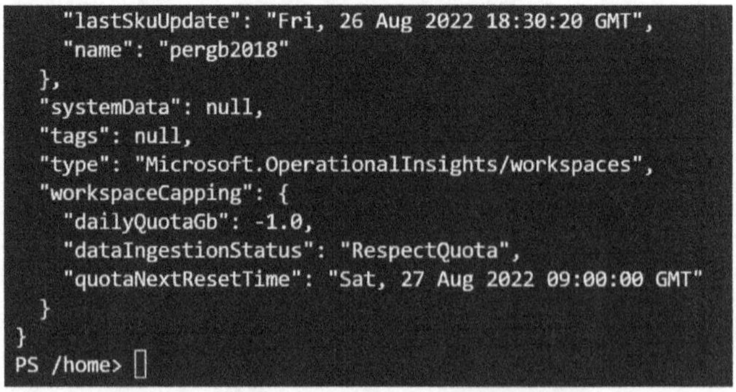
"lastSkuUpdate": "Fri, 26 Aug 2022 18:30:20 GMT",
"name": "pergb2018"
},
"systemData": null,
"tags": null,
"type": "Microsoft.OperationalInsights/workspaces",
"workspaceCapping": {
"dailyQuotaGb": -1.0,
"dataIngestionStatus": "RespectQuota",
"quotaNextResetTime": "Sat, 27 Aug 2022 09:00:00 GMT"
}
}
PS /home>

Figure 11.2 – Creation of a Log Analytics workspace

With the Log Analytics workspace created, it is time to deploy Container insights. Luckily, Azure has made this easy for us as they have created an AKS add-on just for monitoring. Add-ons are a way to bring extra capabilities to your AKS cluster that are fully supported by Microsoft. When you install an add-on to your AKS cluster, the installation, configuration, and life cycle of the resources are managed by AKS.

The next command will get the workspace resource ID and then use that when enabling the monitoring add-on:

```
$resourceIdMonitoring = az monitor log-analytics workspace
show --workspace-name log-aks-smb --resource-group rg-aks-
smb  --query id -o tsv
        az aks enable-addons `
            --addons monitoring `
            --name aks-smb `
            --resource-group rg-aks-smb `
            --workspace-resource-id $resourceIdMonitoring
```

By running this command, some Kubernetes objects are deployed to your AKS cluster in the form of a DaemonSet. If you list all Pods in your cluster, you will see some that contain `OMS` in the name. They are the monitoring objects.

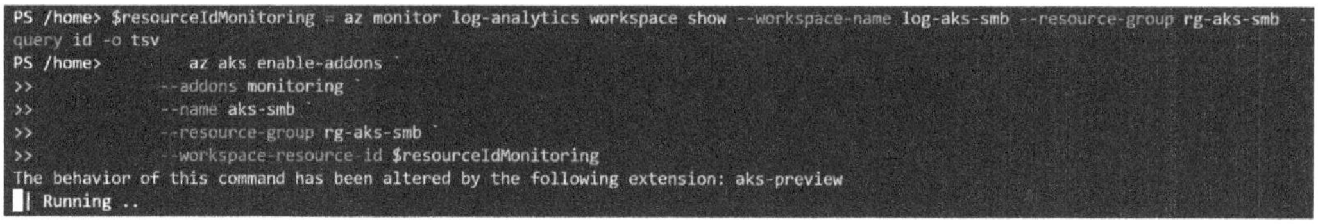
PS /home> $resourceIdMonitoring = az monitor log-analytics workspace show --workspace-name log-aks-smb --resource-group rg-aks-smb --query id -o tsv
PS /home> az aks enable-addons `
>> --addons monitoring `
>> --name aks-smb `
>> --resource-group rg-aks-smb `
>> --workspace-resource-id $resourceIdMonitoring
The behavior of this command has been altered by the following extension: aks-preview
Running ..

Figure 11.3 – Connecting the cluster to Log Analytics

> **Fun Fact**
>
> OMS was the old name for the Log Analytics workspace.

Once this command has finished, your cluster will start sending logs to the Log Analytics workspace. If you jump into the Azure portal and navigate to your AKS cluster, on the left, under **Monitoring**, you will see **Insights**. In the **Insights** blade, you will see not only metrics for the nodes but also your workloads. You can also navigate through and get to the live logs of a running container:

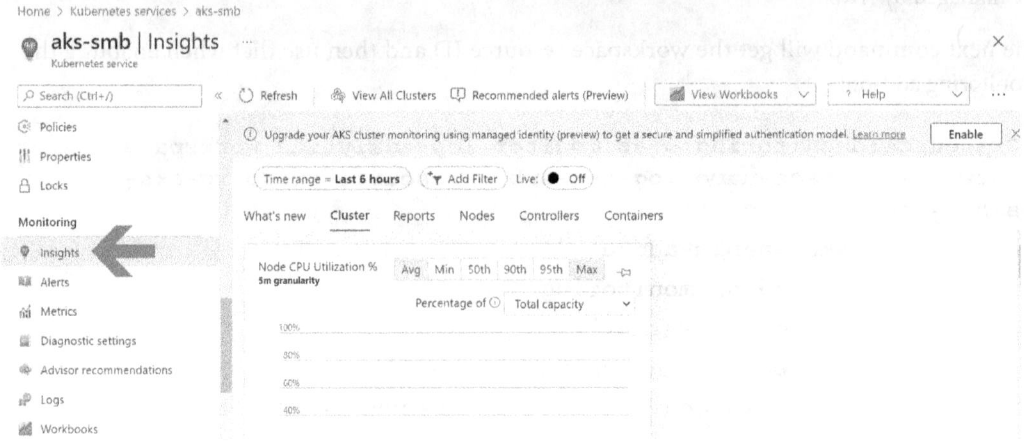
Home > Kubernetes services > aks-smb
aks-smb | Insights
Kubernetes service
Search (Ctrl+/)
Refresh
View All Clusters
Recommended alerts (Preview)
View Workbooks
Help
Policies
Properties
Locks
Monitoring
Insights
Alerts
Metrics
Diagnostic settings
Advisor recommendations
Logs
Workbooks
Upgrade your AKS cluster monitoring using managed identity (preview) to get a secure and simplified authentication model. Learn more
Enable
Time range = Last 6 hours
Add Filter
Live: Off
What's new
Cluster
Reports
Nodes
Controllers
Containers
Node CPU Utilization %
5m granularity
Avg
Min
50th
90th
95th
Max
Percentage of
Total capacity
100%
80%
60%
40%

Figure 11.4 – Container insights enabled

Unfortunately, going over the full capabilities of Container insights is out of the scope of this book, but you can read more at the official docs site at `https://docs.microsoft.com/azure/azure-monitor/containers/container-insights-overview`.

Now we have the monitoring sorted, it's time to look at making the AKS more secure by using the managed identity.

Enabling a managed identity

By default, when you create an AKS cluster, a system-assigned identity will be created for the kubelet. The use case for this managed identity is to authenticate with **Azure Container Registry** (**ARC**) to pull images. You can find this managed identity in the node pool resource group. In this case, the resource group is called `MC_rg-aks-smb_aks-smb_westeurope` and the managed identity is called `aks-smb-agentpool` (`aks-smb` being the cluster name).

You can also have a managed identity for the AKS control plane. By default, this is managed for you as a system-assigned managed identity, but you can bring your own if needed. Why might you want your own? Well basically, if you are using a vNet, Azure disk, static IP, route table, or kubelet user-assigned managed identity outside of your node resource group, Azure is unable to get the principal ID of the system-assigned identity before the cluster is built, and this identity is needed to allow access to the resources outside of the node resource group.

Let's look at enabling the managed identity on our AKS cluster.

We first need to update the cluster to enable the managed identity:

```
az aks update --resource-group rg-aks-smb --name aks-smb
--enable-managed-identity
```

This command can take a while to run depending on the size of your AKS cluster.

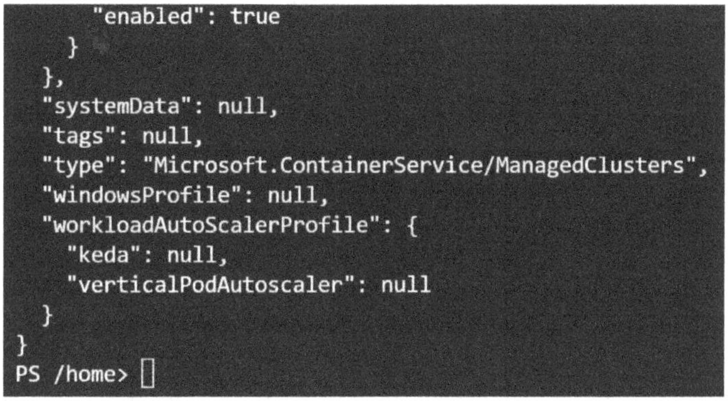
"enabled": true
}
},
"systemData": null,
"tags": null,
"type": "Microsoft.ContainerService/ManagedClusters",
"windowsProfile": null,
"workloadAutoScalerProfile": {
"keda": null,
"verticalPodAutoscaler": null
}
}
PS /home>

Figure 11.5 – Managed identity enabled on the cluster

Now that the cluster can use `managed-identity`, it's time to create our first one. Let's start with the control plane:

```
az identity create --name aks-smb-mid --resource-group rg-aks-
smb
```

Within seconds, the managed identity will be created.

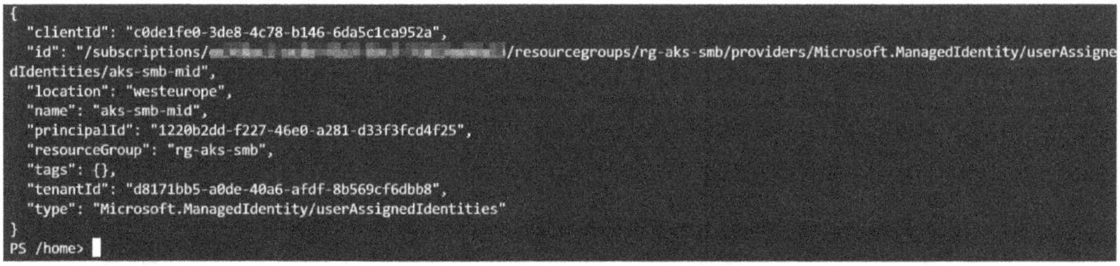
{
"clientId": "c0de1fe0-3de8-4c78-b146-6da5c1ca952a",
"id": "/subscriptions/ /resourcegroups/rg-aks-smb/providers/Microsoft.ManagedIdentity/userAssignedIdentities/aks-smb-mid",
"location": "westeurope",
"name": "aks-smb-mid",
"principalId": "1220b2dd-f227-46e0-a281-d33f3fcd4f25",
"resourceGroup": "rg-aks-smb",
"tags": {},
"tenantId": "d8171bb5-a0de-40a6-afdf-8b569cf6dbb8",
"type": "Microsoft.ManagedIdentity/userAssignedIdentities"
}
PS /home>

Figure 11.6 – Managed identity provisioned

Take a copy of the ID as you will need it for the next command. It's time to update the cluster again.

This time, we assign the newly created managed identity. For the command to work, you will need to add your subscription ID:

```
az aks update --resource-group rg-aks-smb --name
aks-smb --enable-managed-identity --assign-identity /
subscriptions/<subscriptionid>/resourcegroups/rg-aks-smb/
providers/Microsoft.ManagedIdentity/userAssignedIdentities/
aks-smb-mid
```

You will be asked whether you are sure you want to update. Press *Y* on your keyboard to proceed.

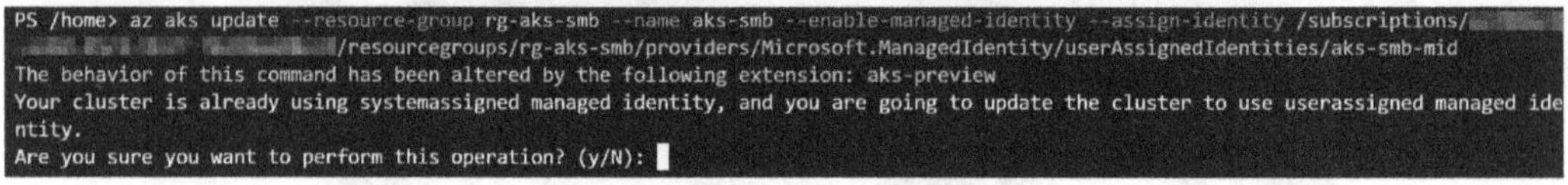
PS /home> az aks update --resource-group rg-aks-smb --name aks-smb --enable-managed-identity --assign-identity /subscriptions/[redacted]/resourcegroups/rg-aks-smb/providers/Microsoft.ManagedIdentity/userAssignedIdentities/aks-smb-mid
The behavior of this command has been altered by the following extension: aks-preview
Your cluster is already using systemassigned managed identity, and you are going to update the cluster to use userassigned managed ide
ntity.
Are you sure you want to perform this operation? (y/N):

Figure 11.7 – Managed identity configured for the control plane

Once this has finished, your cluster will be using a managed identity for the control plane.

> **Important note**
>
> If you are using resources in a resource group that is not your node resource group, make sure you have assigned this managed identity the correct permissions before you perform the update. Check this link for more information on how to set the correct permissions: `https://docs.microsoft.com/azure/aks/use-managed-identity#add-role-assignment-for-control-plane-identity`.

With the control plane managed identity in place, it's now time to add a new managed identity for the kubelet. As with the control plane, we need to create one:

```
Az identity create -name aks-smb-kubelet-mid -resource-group
rg-aks-smb
```

Again, within seconds, the managed identity will be created.

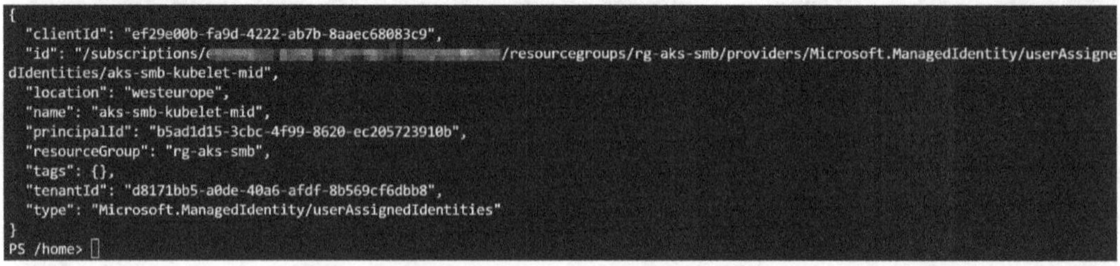
{
"clientId": "ef29e00b-fa9d-4222-ab7b-8aaec68083c9",
"id": "/subscriptions/[redacted]/resourcegroups/rg-aks-smb/providers/Microsoft.ManagedIdentity/userAssignedIdentities/aks-smb-kubelet-mid",
"location": "westeurope",
"name": "aks-smb-kubelet-mid",
"principalId": "b5ad1d15-3cbc-4f99-8620-ec205723910b",
"resourceGroup": "rg-aks-smb",
"tags": {},
"tenantId": "d8171bb5-a0de-40a6-afdf-8b569cf6dbb8",
"type": "Microsoft.ManagedIdentity/userAssignedIdentities"
}
PS /home>

Figure 11.8 – Managed identity provisioned

As with the control plane identity, you will need this ID too, so take a copy.

It's time to update the AKS cluster again. You will notice it is the same command as with the control plane identity, but we are now adding in `-assign-kubelet-identity`. In the real world, you will only need to run this last command, but it's good to understand the process:

```
Az aks update -resource-group rg-aks-smb -name
aks-smb -enable-managed-identity -assign-identity /
subscriptions/<subscriptionid>/resourcegroups/
rg-aks-smb/providers/Microsoft.ManagedIdentity/
userAssignedIdentities/aks-smb-mid -assign-kubelet-identity /
subscriptions/<subscriptionid>/resourcegroups/rg-aks-smb/
providers/Microsoft.ManagedIdentity/userAssignedIdentities/
aks-smb-kubelet-mid
```

As we are updating the kubelet identity that runs on every node, all nodes will need to be upgraded, so depending on the size of your cluster, this could take some time.

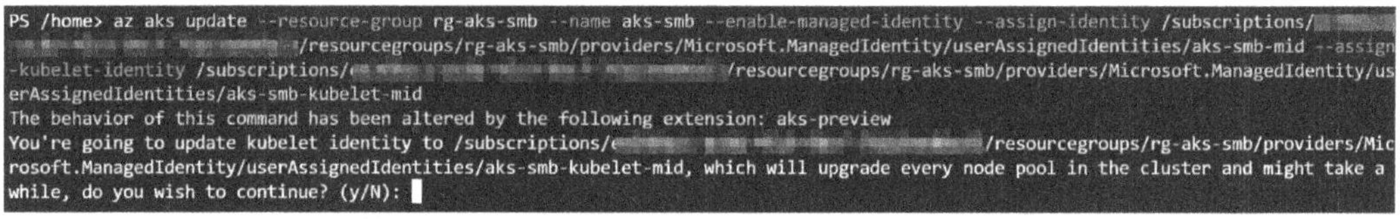
PS /home> az aks update --resource-group rg-aks-smb --name aks-smb --enable-managed-identity --assign-identity /subscriptions/
/resourcegroups/rg-aks-smb/providers/Microsoft.ManagedIdentity/userAssignedIdentities/aks-smb-mid --assign
-kubelet-identity /subscriptions/
/resourcegroups/rg-aks-smb/providers/Microsoft.ManagedIdentity/us
erAssignedIdentities/aks-smb-kubelet-mid
The behavior of this command has been altered by the following extension: aks-preview
You're going to update kubelet identity to /subscriptions/
/resourcegroups/rg-aks-smb/providers/Mic
rosoft.ManagedIdentity/userAssignedIdentities/aks-smb-kubelet-mid, which will upgrade every node pool in the cluster and might take a
while, do you wish to continue? (y/N):

Figure 11.9 – Managed identity configured for kubelet

After a while, your cluster will now be using a managed identity for both the control plane and the kubelet. These managed identities can now be used to communicate with other Azure resources.

With that, it's now time to look at RBAC and securing access to the cluster.

Enabling Azure AD authentication and RBAC

To configure Azure AD authentication and RBAC, we first need to prepare our Azure estate for that. You will need an Azure AD group first. You could use a user, but it is recommended to use a group:

```
az ad group create --display-name aks-smb-admin --mail-nickname
aks-smb-admin
```

Within seconds, the group will be created. You can now add anyone you want into the group, and once we complete the next command, they will have admin access to the cluster. For now, we will not add anyone.

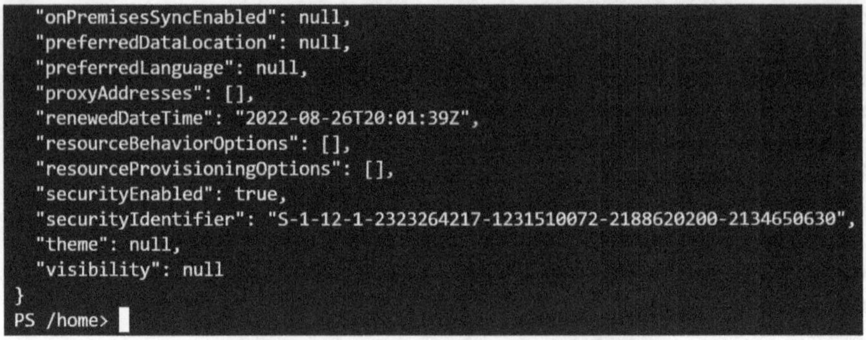
"onPremisesSyncEnabled": null,
"preferredDataLocation": null,
"preferredLanguage": null,
"proxyAddresses": [],
"renewedDateTime": "2022-08-26T20:01:39Z",
"resourceBehaviorOptions": [],
"resourceProvisioningOptions": [],
"securityEnabled": true,
"securityIdentifier": "S-1-12-1-2323264217-1231510072-2188620200-2134650630",
"theme": null,
"visibility": null
}
PS /home>

Figure 11.10 – Creation of an Azure AD group

With the group created, we need to get its ID and use it to upgrade the cluster again. The following command will get the group ID, enable Azure AD integration, and assign the group ID to the cluster-admin role inside the cluster:

```
$groupID = az ad group show --group aks-smb-admin --query id
--output tsv
az aks update --resource-group rg-aks-smb --name aks-smb
--enable-aad --aad-admin-group-object-ids $groupID
```

As with all the AKS update commands we have run in this chapter, this can take some time to complete.

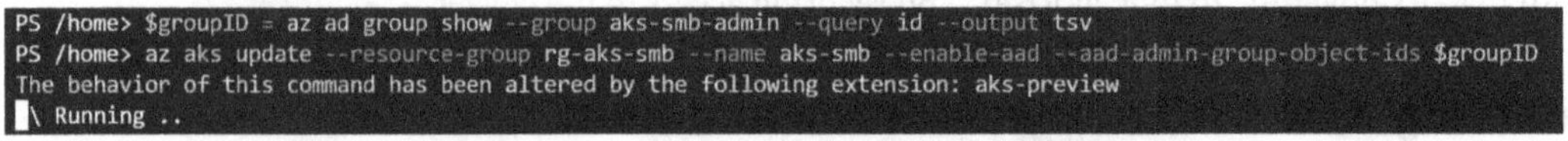
PS /home> $groupID = az ad group show --group aks-smb-admin --query id --output tsv
PS /home> az aks update --resource-group rg-aks-smb --name aks-smb --enable-aad --aad-admin-group-object-ids $groupID
The behavior of this command has been altered by the following extension: aks-preview
\ Running ..

Figure 11.11 – Azure AD authentication configured for the cluster

Now that we have integrated Azure AD, we can enable Azure RBAC. This is nice and easy – just another cluster update:

```
az aks update --resource-group rg-aks-smb --name aks-smb
--enable-azure-rbac
```

Again, with any AKS `update` command, it will take a short while.

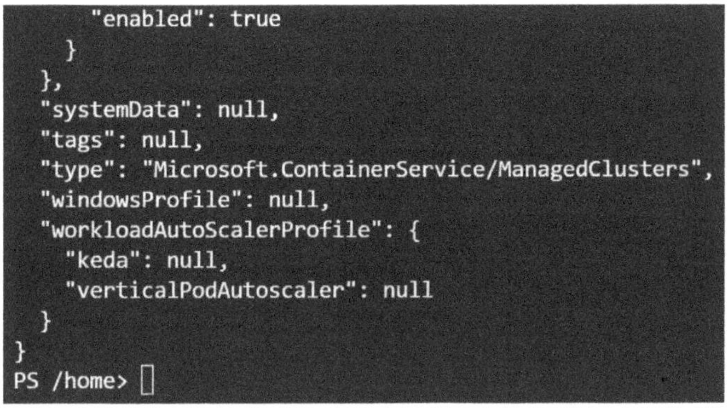
"enabled": true
}
},
"systemData": null,
"tags": null,
"type": "Microsoft.ContainerService/ManagedClusters",
"windowsProfile": null,
"workloadAutoScalerProfile": {
"keda": null,
"verticalPodAutoscaler": null
}
}
PS /home>

Figure 11.12 – RBAC enabled for the cluster

We are now ready to assign ourselves an Azure RBAC role to allow us access to the cluster. For this example, we will use the *Azure Kubernetes Service RBAC Cluster Admin* role to keep things simple. There are other roles you can choose, or you can build your own custom roles. You can read more about the roles at `https://docs.microsoft.com/azure/aks/manage-azure-rbac#create-role-assignments-for-users-to-access-cluster`.

We need a few IDs to be able to assign this permission. First off, our ID, and then the resource ID of the AKS cluster. Once we have both, we can then assign the permissions. The following command will get all this information for you and assign you to that role:

```
$myId = az ad signed-in-user show --query id -o tsv
$aksId = az aks show --resource-group rg-aks-smb --name aks-smb
--query id -o tsv
az role assignment create --role "Azure Kubernetes Service RBAC
Admin" --assignee $myId --scope $aksId
```

A few seconds more and the role is assigned.

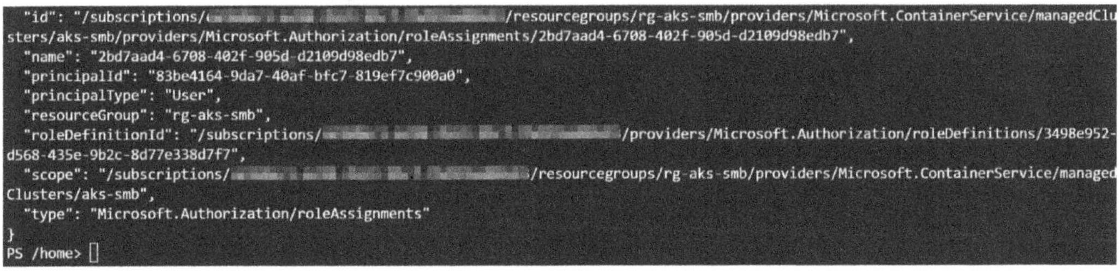
"id": "/subscriptions/ /resourcegroups/rg-aks-smb/providers/Microsoft.ContainerService/managedClusters/aks-smb/providers/Microsoft.Authorization/roleAssignments/2bd7aad4-6708-402f-905d-d2109d98edb7",
"name": "2bd7aad4-6708-402f-905d-d2109d98edb7",
"principalId": "83be4164-9da7-40af-bfc7-819ef7c900a0",
"principalType": "User",
"resourceGroup": "rg-aks-smb",
"roleDefinitionId": "/subscriptions/ /providers/Microsoft.Authorization/roleDefinitions/3498e952-d568-435e-9b2c-8d77e338d7f7",
"scope": "/subscriptions/ /resourcegroups/rg-aks-smb/providers/Microsoft.ContainerService/managedClusters/aks-smb",
"type": "Microsoft.Authorization/roleAssignments"
}
PS /home>

Figure 11.13 – RBAC admin role assigned

We are now ready to remove the local admin account. To do that, we use the following command and, yes, it is another `update` command:

```
az aks update --resource-group rg-aks-smb --name aks-smb
--disable-local-accounts
```

As we know, with an `update` command, it will take a short while.

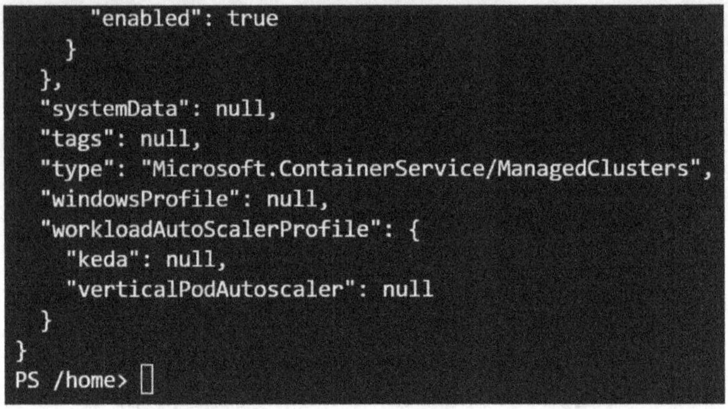
"enabled": true
}
},
"systemData": null,
"tags": null,
"type": "Microsoft.ContainerService/ManagedClusters",
"windowsProfile": null,
"workloadAutoScalerProfile": {
"keda": null,
"verticalPodAutoscaler": null
}
}
PS /home>

Figure 11.14 – Local Kubernetes accounts disabled

Get the cluster credentials again by using the following command:

```
az aks get-credentials --resource-group rg-aks-smb --name
aks-smb
```

Allow the overwrite to update the local copy of `KubeConfig`:

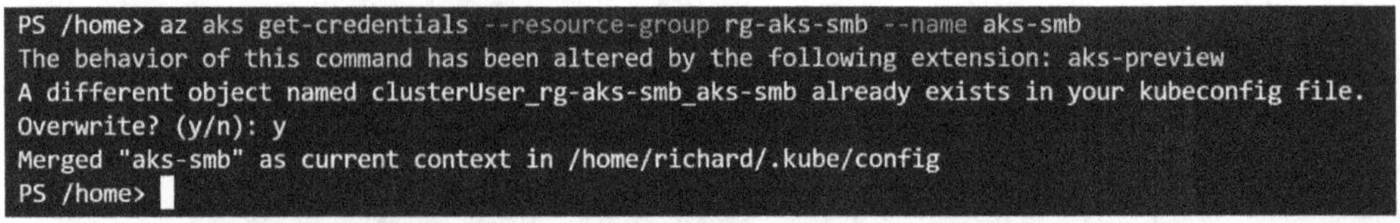
PS /home> az aks get-credentials --resource-group rg-aks-smb --name aks-smb
The behavior of this command has been altered by the following extension: aks-preview
A different object named clusterUser_rg-aks-smb_aks-smb already exists in your kubeconfig file.
Overwrite? (y/n): y
Merged "aks-smb" as current context in /home/richard/.kube/config
PS /home>

Figure 11.15 – Retrieving the credentials for our cluster

Now, it may seem as if you have successfully logged in. But what we technically did was retrieve the `KubeConfig` file.

If we try to get a list of the running Pods, we will be greeted with a message to log in:

```
Kubectl get pods -A
```

We said that just `KubeConfig` wouldn't be enough. After executing this command, we are confronted with a message telling us we can't get a list of all Pods just yet!

```
PS /home> kubectl get pods -A
W0826 20:35:27.968217     728 azure.go:92] WARNING: the azure auth plugin is deprecated in v1.22+, unavailable in v1.25+; use https://
github.com/Azure/kubelogin instead.
To learn more, consult https://kubernetes.io/docs/reference/access-authn-authz/authentication/#client-go-credential-plugins
To sign in, use a web browser to open the page https://microsoft.com/devicelogin and enter the code DAFZ9F933 to authenticate.
```

Figure 11.16 – Message stating we have to authenticate to Azure AD

Once the message pops up, you will need to click the link to the device login page if you are using cloud shell. If you are running the commands locally, you will be confronted with a browser pointing to the Azure AD login page. Follow the instructions (including MFA if you have that set up) and you will successfully log in to the cluster:

```
kube-system    csi-azurefile-node-x8wn8                  3/3   Running   0   38m
kube-system    konnectivity-agent-7b98669647-trscl       1/1   Running   0   38m
kube-system    konnectivity-agent-7b98669647-w8bct       1/1   Running   0   40m
kube-system    kube-proxy-clp8k                          1/1   Running   0   40m
kube-system    kube-proxy-gbrdh                          1/1   Running   0   38m
kube-system    kube-proxy-nqqbt                          1/1   Running   0   44m
kube-system    metrics-server-948cff58d-8h7fw            1/1   Running   0   38m
kube-system    metrics-server-948cff58d-gtn2l            1/1   Running   0   42m
kube-system    omsagent-bgbwz                            2/2   Running   0   44m
kube-system    omsagent-rm9xz                            2/2   Running   0   40m
kube-system    omsagent-rs-6cc5d8ccb-t985t               1/1   Running   0   38m
kube-system    omsagent-tg79k                            2/2   Running   0   38m
PS /home>
```

Figure 11.17 – Successfully logged in

> **Important note**
>
> Please know that you don't have to perform the authentication process for every command. The availability of access to the cluster is dependent on the configuration of your Azure Active Directory (tokens).

Because we also disabled the local admin account, we want to test whether that is really the case. To check whether the local admin account has been disabled, use the following command:

```
az aks get-credentials --resource-group rg-aks-smb --name
aks-smb --admin
```

If the local admin account is successfully disabled, you should get a warning stating you are not allowed to log in using those credentials:

```
PS /home> az aks get-credentials --resource-group rg-aks-smb --name aks-smb --admin
The behavior of this command has been altered by the following extension: aks-preview
(BadRequest) Getting static credential is not allowed because this cluster is set to disable local accounts.
Code: BadRequest
Message: Getting static credential is not allowed because this cluster is set to disable local accounts.
PS /home>
```

Figure 11.18 – Local Kubernetes admin access denied

Now, our cluster is integrated with Azure AD, and we can control Kubernetes RBAC via Azure too. By managing your Kubernetes users via Azure RBAC, you can ensure that if someone leaves your business and you remove the user from Azure, they are also removed from Kubernetes. This highly improves the security of your AKS cluster as you don't have to remember to remove the user from Kubernetes directly. Another cool feature of using Azure RBAC for Kubernetes is that you can use MFA.

With the cluster access secure, it's time to move on to securing the workloads in the cluster. If we look at our scenario, we have the requirement to secure the connection string by storing it in Key Vault and injecting it into the Pod.

Enabling the Key Vault Container Storage Interface

Integration with Azure Key Vault is super simple now thanks to, you guessed it, an AKS add-on! As with the monitoring add-on, when you install the `azure-keyvault-secrets-provider` add-on to your AKS cluster, the installation, configuration, and lifecycle of the resources are managed by AKS.

The more difficult part comes when you want to use a secret from the key vault. This is where the Kubernetes object called a **secret provider class** comes in. We will go into more detail regarding the secret provider class object later in this chapter.

Let's start by looking at enabling the AKS add-on:

```
az aks enable-addons `
    --addons azure-keyvault-secrets-provider `
    --name aks-smb `
    --resource-group rg-aks-smb
```

Running this command can take a few moments as it pulls down and starts up the containers needed for this add-on to work.

```
      "enabled": true
    }
  },
  "systemData": null,
  "tags": null,
  "type": "Microsoft.ContainerService/ManagedClusters",
  "windowsProfile": null,
  "workloadAutoScalerProfile": {
    "keda": null,
    "verticalPodAutoscaler": null
  }
}
PS /home>
```

Figure 11.19 – Enabling the Azure Key Vault CSI add-on

To confirm everything is installed and running, you can use the following command:

```
kubectl get pods -n kube-system -l 'app in (secrets-store-csi-
driver, secrets-store-provider-azure)'
```

You should see a list of Pods all with the status of `Running`:

```
PS /home> kubectl get pods -n kube-system -l 'app in (secrets-store-csi-driver, secrets-store-provider-azure)'
W0828 15:57:09.638680     182 azure.go:92] WARNING: the azure auth plugin is deprecated in v1.22+, unavailable in v1.25+; use https://
github.com/Azure/kubelogin instead.
To learn more, consult https://kubernetes.io/docs/reference/access-authn-authz/authentication/#client-go-credential-plugins
NAME                                     READY   STATUS    RESTARTS   AGE
aks-secrets-store-csi-driver-74d9g       3/3     Running   0          105s
aks-secrets-store-csi-driver-n4ttr       3/3     Running   0          105s
aks-secrets-store-csi-driver-xl8p2       3/3     Running   0          105s
aks-secrets-store-provider-azure-4x2xn   1/1     Running   0          105s
aks-secrets-store-provider-azure-jgjrf   1/1     Running   0          105s
aks-secrets-store-provider-azure-vv7gb   1/1     Running   0          105s
PS /home>
```

Figure 11.20 – Verifying that Azure Key Vault CSI Pods are running

We now need to create a key vault to store the connection string into. You could use an existing one if you have one, but for this demo, we will create one with the following command:

```
az keyvault create --name kv-smb --resource-group rg-aks-smb
--location westeurope
```

> **Important note**
>
> Key vault names need to be globally unique, so if and when you try to create the key vault using the preceding command and it errors, just change the name to something you can remember, as you will need it later in this chapter.

After a couple of seconds or so, the key vault will be created:

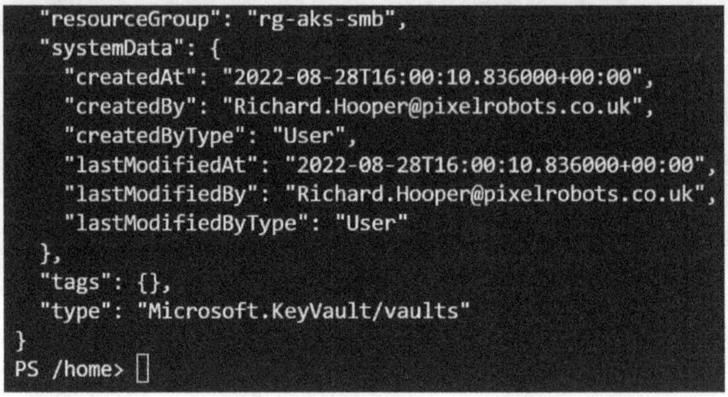
"resourceGroup": "rg-aks-smb",
"systemData": {
"createdAt": "2022-08-28T16:00:10.836000+00:00",
"createdBy": "Richard.Hooper@pixelrobots.co.uk",
"createdByType": "User",
"lastModifiedAt": "2022-08-28T16:00:10.836000+00:00",
"lastModifiedBy": "Richard.Hooper@pixelrobots.co.uk",
"lastModifiedByType": "User"
},
"tags": {},
"type": "Microsoft.KeyVault/vaults"
}
PS /home>

Figure 11.21 – Azure Key Vault creation

With the key vault created, we are ready to set permissions on it to allow the AKS cluster to pull secrets. For this, we will be using a managed identity. You could use one of the ones you created earlier, or you could use the one that gets created for you when you enable the AKS add-on. For now, we will use the one that was created for us. You will find it in the node resource group with the name `azurekeyvaultsecretsprovider-aks-smb` (`aks-smb` being the cluster name).

The following command will get the client ID of the managed identity and will also set the `get` secret permission on the key vault:

```
$MIClientId = az aks show -g rg-aks-smb -n aks-smb --query
addonProfiles.azureKeyvaultSecretsProvider.identity.clientId -o
tsv
 az keyvault set-policy --name kv-smb --secret-permissions get
--spn $MIClientId
```

After this command completes, the managed identity will have access to get any secrets in the key vault:

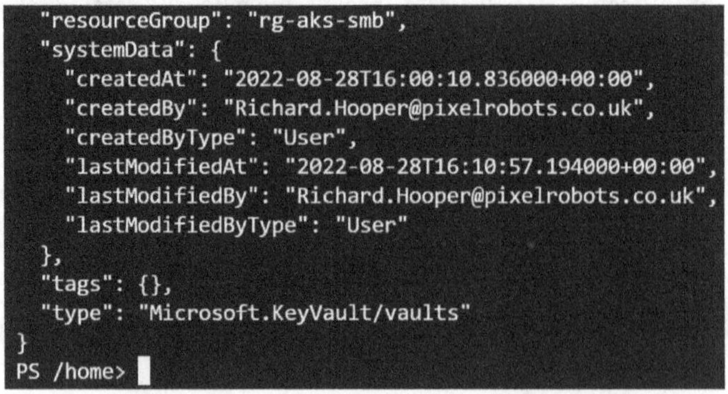
"resourceGroup": "rg-aks-smb",
"systemData": {
"createdAt": "2022-08-28T16:00:10.836000+00:00",
"createdBy": "Richard.Hooper@pixelrobots.co.uk",
"createdByType": "User",
"lastModifiedAt": "2022-08-28T16:10:57.194000+00:00",
"lastModifiedBy": "Richard.Hooper@pixelrobots.co.uk",
"lastModifiedByType": "User"
},
"tags": {},
"type": "Microsoft.KeyVault/vaults"
}
PS /home>

Figure 11.22 – Key Vault policy configured

Next, we need a secret in the key vault. This will be the connection string for the database. In the real world, you would have this automated when you create a SQL database, but for ease, we will just create a dummy secret now:

```
az keyvault secret set --vault-name kv-smb -n customer1-
connectionstring --value connectionstring
```

Within an instance, the secret is created.

```
    "updated": "2022-08-28T16:15:43+00:00"
  },
  "contentType": null,
  "id": "https://kv-smb.vault.azure.net/secrets/customer1-connectionstring/32f89886c4ea45b09806d74937d5f4a8",
  "kid": null,
  "managed": null,
  "name": "customer1-connectionstring",
  "tags": {
    "file-encoding": "utf-8"
  },
  "value": "connectionstring"
}
PS /home> 
```

Figure 11.23 – Key Vault secret created

We are now on to the secret provider class. This Kubernetes object tells the cluster where the key vault is, what managed identity to use, and what tenant the managed identity is in. You will need to fill in any values inside <>. You can find the secret provider class manifest in the Git repo here: `https://github.com/PacktPublishing/Azure-Containers-Explained/blob/main/CH11/SecretProviderClass.yaml`.

The most important part of this file is the `object` section. Here is where you set what secrets you want your Pods to be able to see from the key vault. We are only providing one: the connection string for `customer1`. You will need to deploy a secret provider class per customer in our scenario, so you will need to update `objectName` to match also.

Once you have updated your file with your `client-id`, `key-vault-name`, and `tenant-id` values, save the file and apply it using `kubectl`:

```
kubectl apply -f SecretProviderClass.yaml
```

If you have updated the file correctly, you should now have your secret provider class created:

```
PS /home/richard> kubectl apply -f SecretProviderClass.yaml
W0828 16:28:42.127813     374 azure.go:92] WARNING: the azure auth plugin is deprecated in v1.22+, unavailable in v1.25+; use https://
github.com/Azure/kubelogin instead.
To learn more, consult https://kubernetes.io/docs/reference/access-authn-authz/authentication/#client-go-credential-plugins
secretproviderclass.secrets-store.csi.x-k8s.io/azure-kvname-user-msi created
PS /home/richard> 
```

Figure 11.24 – Secret provider class for the key vault secret configured

With the secret provider class created, it is finally time to reapply the frontend deployment manifest but with a few additions: `volumes` and `volumeMounts`.

Here is a snippet of the extra bits we are adding to the manifest:

```
        volumeMounts:
        - name: secrets-store-inline
          mountPath: "/mnt/secrets-store"
          readOnly: true
    volumes:
      - name: secrets-store-inline
        csi:
          driver: secrets-store.csi.k8s.io
          readOnly: true
          volumeAttributes:
            secretProviderClass: "customer1-secrets"
```

What this is doing is mounting any of the objects we referenced in the secret provider class into the `/mnt/secrets-store` location inside the Pods in the deployment. The last line tells Kubernetes which secret provider class to use. We can use `kubectl` to apply this update to our existing deployment:

```
kubectl apply -f https://raw.githubusercontent.com/
PacktPublishing/-Up-and-Running-with-Azure-Containers/main/
CH11/frontend.yaml
```

You will notice that only the deployment has been configured:

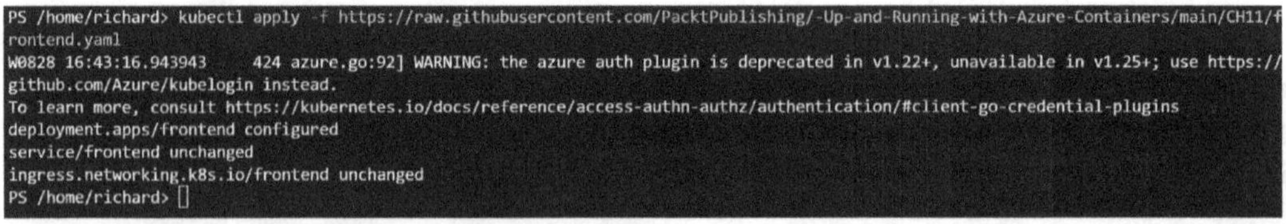
PS /home/richard> kubectl apply -f https://raw.githubusercontent.com/PacktPublishing/-Up-and-Running-with-Azure-Containers/main/CH11/frontend.yaml
W0828 16:43:16.943943 424 azure.go:92] WARNING: the azure auth plugin is deprecated in v1.22+, unavailable in v1.25+; use https://github.com/Azure/kubelogin instead.
To learn more, consult https://kubernetes.io/docs/reference/access-authn-authz/authentication/#client-go-credential-plugins
deployment.apps/frontend configured
service/frontend unchanged
ingress.networking.k8s.io/frontend unchanged
PS /home/richard>

Figure 11.25 – Frontend deployed

To check whether the secret is now present inside the Pod, we can use the `kubectl exec` command. But first, we need the name of the frontend Pod:

```
kubectl get pods
```

Once we have the Pod name, we will then go ahead and reveal the secret using `kubectl exec` and the `cat` command:

```
kubectl exec frontend-cffdb55c5-x8g8f -- cat /mnt/secrets-
store/customer1-connectionstring
```

The last line in the PowerShell prompt is the secret:

```
PS /home/richard> kubectl get pods
W0828 16:50:28.813103     544 azure.go:92] WARNING: the azure auth plugin is deprecated in v1.22+, unavailable in v1.25+; use https://
github.com/Azure/kubelogin instead.
To learn more, consult https://kubernetes.io/docs/reference/access-authn-authz/authentication/#client-go-credential-plugins
NAME                              READY   STATUS    RESTARTS   AGE
billingstatements-74cb855d74-z6czb   1/1     Running   0          44h
frontend-cffdb55c5-x8g8f          1/1     Running   0          115s
PS /home/richard> kubectl exec frontend-cffdb55c5-x8g8f -- cat /mnt/secrets-store/customer1-connectionstring
W0828 16:50:35.368885     548 azure.go:92] WARNING: the azure auth plugin is deprecated in v1.22+, unavailable in v1.25+; use https://
github.com/Azure/kubelogin instead.
To learn more, consult https://kubernetes.io/docs/reference/access-authn-authz/authentication/#client-go-credential-plugins
connectionstring
PS /home/richard>
```

Figure 11.26 – Secret available inside the Pod

Now we just need to ensure that our code can read this newly mounted secret. Our workloads are now more secure as they reference key vaults for our secrets and the secrets are only stored within the Pod.

We have now successfully configured several best practices on AKS without having to redeploy our cluster. Please know you can configure all of these from the start. However, that's not always the way it goes.

Learnings and future decisions

AKS has grown in the past years. Where a lot of features we have been using throughout this book would traditionally require a complete redeployment of the cluster, these days, they do not. That makes Azure Kubernetes a reliable platform for an enterprise (customer) to run their containerized solutions on. Not that Kubernetes is not reliable by default, but it is very hard to explain to your customers that there is going to be downtime because you want to add a feature to your platform.

Enterprises care about governance and security and, as we stated in the introduction, if you're starting small, you might have not thought about all of it. And if there is anything to be learned from this chapter, maintaining the quality of your AKS platform is a continuous process. All the things we have configured in the past two chapters merely scratch the surface of what you can actually do and what is to come. As long as you are willing to add or discard features depending on whether you and your customers need them, you will be set for a long time to come.

That brings us to another lesson. By configuring these features, we have deployed quite a lot of Pods to the AKS cluster. This adds overhead to the environment. They are needed but we also want to minimize overhead. Don't be afraid to delete features you are no longer using. Even though these Pods seem like small entities, they provide services to everything in the cluster and might even require significant resources. If you don't need them, get rid of them.

Oh… and the biggest lesson that we didn't discuss here? Deploy new configuration to a testing environment first! You don't want to accidentally bring down the cluster or lock yourself out!

Summary

In this chapter, we have learned that we can build upon an existing AKS cluster and add features without having to redeploy the environment. We have also learned that enterprise customers require more from you – not because they are annoying, but because the bigger the customer, the bigger the impact once things go south. We want to mitigate those risks but also provide a platform that is operable by our Ops teams.

With the Azure Monitoring plugin for Container insights, we have also seen that it is quite easy and effortless to configure monitoring for your containers. It is a no-brainer to configure but often left out.

Everything in this chapter is focused on two things:

- Providing the features and configuration your customers require
- Lowering the overhead and maintenance for your Ops teams

We have been focusing on the first bullet. But if you were to replace the use case and focus on lowering effort and management from your Ops team, almost the same configuration applies. To summarize, there can be only winners.

In the next chapter, we will recap what we have implemented based on our use cases so far and summarize the technical constraints we have introduced while doing so.

Part 3: Migrating Between Technologies and Beyond

Whether you are a start-up, a small or medium-sized business, or an enterprise, requirements change over time. To meet new requirements, new technologies often need to be introduced. In the third part of this book, we will look at the origin technology and explain what options you have for migrating to a different technology. We will look at what this requires and what changes will need to be made, but also what value you will add by doing so.

This part comprises the following chapters:

- *Chapter 12, Migrating Between Container Technologies*
- *Chapter 13, Azure Container Instances – I Like the Scalability, But I Need More*
- *Chapter 14, Beyond Azure Container Apps*
- *Chapter 15, Azure Kubernetes Service – The Next Steps*
- *Chapter 16, What's Next – Diving into Future Trends and More*

12
Migrating Between Container Technologies

In *Part 1, Understanding the Azure Container Technologies*, and *Part 2, Choosing and Applying the Right Technology*, we took a dive into the different Microsoft Azure technologies, and which use cases are applicable to them, both on a technical and business level.

As we have stated multiple times throughout this book, business requirements change. Therefore, your use case may change and the technology you initially selected as your preferred Azure container technology is no longer the best technical fit.

This may be challenging for your Ops department, but it is a good thing! It means you're advancing, maybe selling more, and acquiring bigger customers. Or maybe you were a start-up and just selected the most cost-efficient Azure container technology to get started. There is no shame in revisiting past choices. If anything, remaining on the same technology for years on a public cloud might very well mean that you are standing still and have stopped adapting.

In *Part 3* of this book, we will look at migrating between different technologies, why you would and should migrate, and what is required to successfully perform that migration. We have shared a lot of technical examples in the past chapters, and you could argue that using them would not be enough to successfully redeploy to new technology. And to some degree, that would be correct. However, there is a thought process behind it. Simply moving things around is not going to help you satisfy the business needs, whether that is in terms of governance, costs, or specific features.

In this chapter, we're going to go over the following topics:

- Revisiting technologies and useability
- Investigating use cases for migrating between technologies
- Exploring migration paths

Revisiting technologies

Over the course of the past chapters, we have discussed five different Azure container technologies:

- Containers for Azure Functions
- Containers on Azure App Services
- Azure Container Instances
- Azure Container Apps
- Azure Kubernetes Service

First things first. All these technologies have different use cases for running containerized solutions on Microsoft Azure. However, we are going to completely forget about Azure Functions for now. As the use case for containers on Azure Functions is to overcome technical blockers on Azure Functions itself (for example, an unsupported runtime), it is rarely the go-to resource for initially deploying your containerized solution.

The same could be said for using Azure App Services but there is a little more to that. Let's briefly revisit these technologies and their use cases.

Azure App Services

Containers can run on Azure App Services. They run very well and a single container deployment is a very common scenario. We use Azure App Services to host our containers if we are already using Azure App Services or we are used to their *way of operating.*

Alternatively, a big use case for using Azure App Services to run your containers is pricing. Azure App Services is relatively cheap if you compare its competitors: Azure Containers Apps and **Azure Kubernetes Services** (**AKS**). You could argue that **Azure Container Instances** (**ACI**) would be a good challenger; however, we mainly use ACI for burstable and on-demand workloads. If you have written a frontend application that is being accessed continuously, we are left with three technologies: Azure App Services, Azure Container Apps, and AKS.

Azure Container Apps itself is still relatively new so chances are you already had things up and running in Azure App Services before May 2022 when Azure Container Apps was released to production.

Now, if we just wanted to run a couple of containerized frontends before May 2022, you would need to do some serious convincing to your Chief Financial Officer as Azure App Services is a lot cheaper than AKS.

But the biggest reason for using Azure App Services that we have seen across customers is the familiarity with Azure App Services and the trust they have in the platform, and now that they started using containerized solutions, it simply sounded like the logical choice.

We can run single and multi-container (Docker Compose) workloads on Azure App Services. Truth be told, you will miss a lot of Azure App Services features when running containers but the question is whether you really need those for containers in general.

So, we can summarize the Azure App Services use case as follows:

- Cheap compared to alternatives
- Single and multi-container support
- Familiarity with the platform
- Works great if you don't need a lot of container-specific features

> **Important note**
>
> During the course of the next chapters, we will not go into detail on how to move from Azure App Services to other Azure technologies. As App Services only requires a Dockerfile or container image to run the container, there are no other *moving* parts we need to take into account. You can simply pick another technology.

It is important to note that Azure App Services was not originally designed to run containers. It was simply a feature that was added at some point in time.

Azure Container Instances

Azure Container Instances are perfect for burstable and on-demand workloads. Plenty of software architectures have some kind of job or on-demand workload running somewhere. For containers, ACI is your go-to resource.

You would usually select ACI as your preferred technology when you only have a few containers and you want them to run only when you need them. ACI doesn't provide a whole lot of features but the important features such as network integration, scalability, and speed are definitely there.

Across customers, we usually see Azure containers only when all other parts of the solution are not containerized and are running on another **Platform as Service** (**PaaS**) technology such as Azure App Services, Azure SQL, or even IaaS/virtual machines.

Additionally, ACI can be used as the burstable platform for your AKS. We call this **Virtual Kubelet**. Let's say everything we have is running on AKS, but we do some batch processing that is taking up too many resources on the AKS cluster. We simply don't want to trigger our autoscaling and add another node and pay for it when something or someone requires that batch process to start. This is when ACI is the perfect addition to your AKS cluster.

So why Container Instances? These are the advantages:

- Easy to deploy
- Fast
- Very well suited for scalable workloads
- Basic integration with Azure features

Azure Container Apps

The new player on the block is Azure Container Apps. As of May 2022, Azure Container Apps is "generally available," meaning Microsoft is officially supporting it for production workloads.

We usually say and hear the following: Azure Container Apps is basically Kubernetes without the complex platform management. However, that doesn't do Azure Container Apps the justice it deserves and doesn't show Kubernetes the respect it deserves as a platform. In fact, they are two different platforms with different use cases.

Azure Container Apps has a fairly simple deployment model and doesn't require the learning investment required for AKS from a Dev and Ops perspective. That is very true. However, it also has constraints. You could describe Azure Container Apps as the SaaS version of AKS. And with SaaS comes constraints and less flexibility than a platform you can configure to your liking.

Azure Containers Apps comes with Dapr and KEDA built in. That may not sound like much but if you are building a solution based on a microservices architecture and want to use container technologies to run them, Dapr and KEDA are two technologies you want to incorporate into your best practices.

Building is the keyword here. If you are building something new and it is containerized and you want to follow the best practices and be ready for *full-scale* Kubernetes in the future, then Azure Container Apps is the platform for you.

During Microsoft Ignite in 2021, Azure Container Apps was released in preview and was described as "the stepping stone to AKS," and that is something we completely agree with.

Using Azure Container Apps lets your developers focus on building a microservices solution running in containers and deploying them without much knowledge of the platform it is running on. This may sound a bit odd, but it is actually a major use case: we start writing code and we want to deploy it, but we need complex infrastructure for that. This was exactly the problem with AKS (and Kubernetes in general). There were two learning curves: writing and building a new solution, and learning and managing a complex platform such as Kubernetes.

Azure Container Apps is still relatively new, but we expect a lot from it.

You will use Azure Container Apps for the following reasons:

- You are starting from scratch and are building microservices
- As a stepping stone to AKS
- To enforce best practices to use containers the right way
- AKS is not yet fit for your use case

Azure Kubernetes Service

If you run your solutions containerized and your standard is to release new features and services as containers, then this is where you will end up eventually: AKS.

AKS is the Microsoft Azure version of *managed Kubernetes*. In the previous paragraph, we briefly mentioned the complexity of AKS. That complexity is not in its *deployment* but in the Ops side of things. AKS is deployed quite easily and the very complex parts of Kubernetes itself, such as the control plane, the database (etcd), and the configuration between the control plane and nodes, are abstracted away. Microsoft Azure manages that for you.

The complexity comes with implementing best practices, but this is also the beauty of AKS. We can integrate with plenty of Azure technologies such as Azure Key Vault, Azure Container Registry, Azure Active Directory, and many more. But we can also implement third-party solutions such as KEDA, Dapr, GitOps with Flux or ArgoCD, or a multitude of different Ingress controllers.

The freedom in the platform allows us to deploy pretty much whatever we want. The complexity is in managing those best practices and making sure the platform remains secure, cost-efficient, and easy to operate.

You use AKS because of the following reasons:

- You have outgrown your Azure Container Apps
- You want flexibility and are willing to learn about Kubernetes
- You want to integrate with different Azure and third-party technologies
- Governance, security, and compliance are important and your customers start demanding them

To summarize, we can use each technology for different use cases, but the most common ones are as follows:

Platform	When
Azure Functions	To overcome limitations (for example, runtimes) on Azure Functions
Azure App Services	Run single and multi-instance containers for financial reasons and because you already feel very familiar with App Services
Azure Container Instances	Small burstable and on-demand workloads and, or in addition to, AKS
Azure Container Apps	Newly developed solutions or smaller solutions that are not big enough for Kubernetes (yet)
Azure Kubernetes Service	Bigger solutions that run in production and require integrations with solutions such as Azure Key Vault, complex routing through virtual networks, and/or other third-party Kubernetes plugins

Table 12.1 – Summary of Azure container technologies and when to use them

Now, let's talk about use cases. Why would you want to migrate between different container technologies? There are a lot of reasons but let us discuss the most common (and sometimes, not so obvious) ones.

Investigating use cases for migrating between technologies

It is better to make a bad decision than to make no decision at all. Now, let's not assume you made a bad decision and choose the wrong container platform. But if you have never made a decision, you wouldn't have been running containers on Azure in the first place!

What we are simply trying to explain is that a container is a container, and no matter on what platform, it's still a container. Whether you have different requirements that motivate you to move to a different container platform or because a past decision didn't turn out so well, it does not matter. As long as we have the containers, we can talk about which platform to move them to.

We are not going to dive into each technology separately here; that is what the upcoming chapters are for. At some point in time, you have made a decision to go with a specific container technology in Azure. Let's just say that whatever you selected, we assume it was the right choice. But… the world is changing, customers are changing, your company is changing, and so requirements will change. This is inevitable. Don't see migrating between technologies as a failure to select the correct technology from the start, but as an improvement that will make your business better. After all, maybe you started with a single container and AKS was never relevant but now it might just be.

Let us take a look at the common reasons and use cases for migrating between technologies.

Customer requirements and features

If you are a software company, you have a plan for what you want to build for the upcoming year. These plans are often a result of customer input and requirements and your own vision of improving your solution. The product owner prioritizes these and at some point, you will run into the question, *can the platform we use even do this?* Hopefully, you have not just looked at the *current* situation when selecting your initial container platform on Azure but also took the plans you have into account. If that is the case, you can likely plan 6-12 months ahead, and that is plenty of time to migrate!

If you are making the decision to research a new platform for your solution and invest in it, it is important you do this based on the requirements and features you have prioritized in your product backlog and plan for the next 6-12 months. Simply switching platforms because one customer demands a specific feature will send you down a rabbit hole that can be very, very deep.

We often see these changes based on new requirements in governance, networking, and security. For example, if you are using Azure App Services and customers now demand access on a network level to your solution (virtual private networks or vNet peering), you will have to move to a different technology to satisfy those requirements. Not because App Services are not capable of providing these services but because other platforms are more scalable and cost-efficient when it comes to this.

In any case, this is a good thing. It is usually the result of being successful, gaining more customers, and understanding your customer requirements better. If you show that you are adapting to the market, customers are perfectly fine with a migration. If anything, it will show them you are thinking ahead.

Money, money, money

As a software company, you want a couple of things. You want to add as many customers as possible, and you want to add value for your customers, but in the end, you also want to make some money.

You are running a business, paying your employees, having costs simply for doing business, and in the end, you want to make some profit. Want you want is to sell as much as you can and keep costs to a minimum, which includes platform costs.

Migrating between technologies might very well be a financial decision. You want to make sure your platform costs are kept to a minimum but comply with customer requirements and expectations. Perhaps you are providing a subscription model (let's say monthly), and your customers pay you a price per user. If your solution always requires a new App Services plan with app services for each customer, whether that is for 10 or 100 users, smaller customers are not good for margin. You can get away with this if it is a minority of customers and you based your subscription prices on an average. But at some point, things can be more efficient from a financial perspective.

The choice of using Azure App Services was technical and financial: you want single tenancy for security reasons and Azure App Services costs are predictable and not that high. You added a couple of features, resulting in additional App Services in your App Services plan and you ended up in a scenario where each customer gets their own Azure App Services. Now that is becoming very expensive. And believe it or not, maybe you can move this to AKS, which to be fair, is quite the overkill for a single container but is one of the best technologies when it comes to deploying single-tenant solutions in a multi-tenant backend (the AKS cluster). On top of that, running multiple containers makes AKS very efficient in terms of overhead, scalability, and security.

Moving between platforms can be a financial decision. That doesn't always land well with your technical team, but our experience is that as long as you are transparent on the *why*, there is a level of understanding.

Deprecation

Well, we can be very short about this. Technologies can be deprecated. Microsoft announces ahead of time (years ahead sometimes) so it can never be a surprise. Now, when you are using Microsoft Azure, which we assume you are as this is what this book is about, there will always be a new technology to migrate to. It is very uncommon that Microsoft announces the deprecation of a technology without having a new technology readily available. That doesn't mean migrating is going to be a simple thing but there is almost always an alternative. And we are going to be very honest: if you are struck by surprise when you wake up in the morning and the technology you are using is no longer available, you have not been paying attention to what (arguably) your most important partner is doing and what their plans are for quite some time.

Now, the deprecation of an entire technology is not very common. But there can also be features in the technology you are using that will be deprecated. We recommend always (at least monthly) checking up on the Microsoft announcements and blogs to stay informed and help you plan ahead.

New technologies

Public clouds develop fast. Really fast. When it comes to Microsoft Azure, there are new features released every month and there are new technologies being announced during every big conference, such as Microsoft Build and Microsoft Ignite. You will want to stay on top of that.

A new technology may very well mean a new way to deploy and run your solution on Microsoft Azure. That may come with new features that you have been waiting for or new features you have been waiting for but didn't know you were! New technologies also come with their own pricing model. If that new technology contains all the features you need and is much cheaper than what you are currently using, why not?

However, most of the time, you will use new technology because it contains features that you need, or simply because it is easier to deploy. Nine out of ten times, this will be a technological decision. Let's picture this example: in 2021, Azure Container Apps was released in preview. Until then, it was either Azure App Services, ACI (to some degree), or AKS that you could use. To be fair, most container workloads would go to AKS. That was a pretty big gap that Microsoft filled with the introduction of Azure Container Apps.

For smaller and simpler workloads, we were normally pointed to AKS but now we have Azure Container Apps. Moving from AKS to Azure Container Apps isn't something a lot of people think of, but if the use case is technically and financially right for you, why not?

Allocating time for innovation

Now, this is not a use case by itself but something we want to share with you. As we have mentioned, new technologies on the public cloud are released every month, sometimes every week. Cloud development is fast-paced. If your use case changes, you want to be ready for that and invest time. A change in requirements is not a bad thing; it's an opportunity that is both financially and technically beneficial.

Whether you work in a DevOps motion and are very agile or you use any other (or no) framework, allocate time for innovation. Provide your employees with time to research new and different technologies. In fact, don't just do that. Explain to them how your business works and what your long-term vision is. You will be surprised by what engineers come up with. If you do this, migrations between technologies are a piece of cake.

Exploring migration paths

Next comes the big question about migration paths. With the different technologies, what migration paths are we really talking about? You could argue that you can migrate from anything to anything, and yes, that is true, but we are focusing on the most common scenarios. We feel that a lot of companies are stuck with one technology and are afraid to move to another. That is understandable because, in the end, everything you provide to your customers is hosted on that platform. In the next chapters, we will go through the most common scenarios and, hopefully, give you something to grasp when one of the aforementioned use cases becomes a topic in your company.

Let us see what migration paths we are talking about.

Azure App Services to "anything"

We are literally saying to *anything*. As mentioned in earlier paragraphs, Azure App Services only requires a Dockerfile or container image to get started. That means that by using any other technology, you are pretty much starting from scratch for that technology. If you are moving to ACI, you need to invest in writing your infrastructure as code for ACI. If you are moving to AKS, then you still need to write the YAML or Helm charts to run your solution on AKS.

ACI to Azure Container Apps

We find it unlikely to move from ACI back to Azure App Services. This could've been a migration path, but we now have Azure Container Apps. If, for whatever reason, ACI is not living up to your expectations anymore or you are redeveloping a few things, then Azure Container Apps is definitely a technology you can leverage. But, let us not forget AKS.

Both technologies require us to write YAML. Where Azure Container Apps requires some YAML that is leveraged through ARM templates or Azure Bicep, AKS itself requires the YAML that we normally use when deploying Pods to AKS.

If you have been using ACI and went down the YAML route, you still need to refactor a large part of that. However, you do have all the information you need to run your container; it is simply a matter of refactoring.

But let's not forget about the networking integration that is often used in ACI. That will likely be the biggest task during migration to any other technology; customers might be connecting over an Application Gateway or an Azure Front Door. Or maybe you are using a NAT gateway to provide a static outgoing IP. Those are your concerns; the YAML part is pretty straightforward. You can read how we would go about doing this in *Chapter 13*.

Azure Container Apps to AKS

Okay, you were the early adopter and are using Azure Container Apps. It's not very likely that you will change very soon but let's say you will. Also, it is unlikely that you need to move back to ACI. If that was the case, you probably made the wrong decision somewhere or you need to reduce costs. But there is good news: if you have used Azure Container Apps to actually get started with containers and have interpreted the technology as *the stepping stone to AKS*, you are going to be in for a very easy migration.

AKS can run at least the same add-ons/technologies as Azure Container Apps does. That means Ingress, KEDA, and Dapr. They are fairly easy to install by leveraging the right Helm chart or AKS add-on, which makes you good to go. However, Azure Container Apps works with app environments, which act as a secure boundary between different groups of container apps.

Before we can get started with AKS, we really need to think about our setup. How can we deploy something that can at least provide the same security boundaries? The answer to that is namespaces, network policies, and RBAC. We will elaborate on how we go about that in *Chapter 14*.

AKS

So you ended up at AKS as we told you would happen. As AKS provides so many integrations with Azure itself and third-party technologies, adding new features may very well be migration on its own. It is likely that you started with a basic deployment, but now the use case has changed or there are new requirements. When it comes to features and requirements, it is very rare that you have to answer with "no, we can't do that." But implementing those features can be quite a challenge. Sometimes it even requires redeploying the entire cluster and sometimes it requires redeploying current running solutions. In *Chapter 15*, we will dive into how we go about adding new features to AKS and setting up our deployments to make sure we can add those features and redeploy fast when we need them.

Using AKS doesn't mean you are *there* and you made it. It means you are now ready to embrace the fact that the sky is the limit, and you need to think things through.

Summary

In this chapter, we have learned that it is likely you will need to migrate to a different Azure container technology at some point. Whether that's for financial or technical reasons, at some point, it will happen. Your solution has a life cycle and so does your platform configuration; if it is time to move on, then it's time.

We have seen that Azure Functions itself is a very specific use case and it is unlikely you are looking for migration scenarios if you are using those. Azure App Services, on the other hand, is a big use case for migrations, but migrations are fairly simple as they are not really a container platform but more a platform that supports running containers.

Things get interesting when you are running ACI, Azure Container Apps, and AKS, because those are the real container platforms on Azure, and whichever use case is applicable for you, you might just have to migrate between them or use a combination of the two!

13

Azure Container Instances – I Like the Scalability But I Need More

And here we are, in the world of **Azure Container Instances** (**ACI**). We are assuming you have ACI up and running and that this was a great initial choice. However, things change over time, and the logical thing to do is evaluate whether ACI is still the right solution for you. Whether that is from a financial or technical perspective, it might be time to move on.

However, change can be difficult. Deploying your container onto another platform initially sounds very simple. After all, it's just a container. However, after the initial excitement of a successful deployment comes the realization that features in other technologies can be fundamentally different from what ACI provides. There are two ways to go about this. We can either follow the trial-and-error route until everything is up and running, or we can think things through, research, and make a plan.

We think it's obvious that, for production scenarios, everyone favors the latter. And that is exactly what this chapter is about – thinking things through and determining how we can move from ACI to another technology.

In this chapter, we're going to cover the following topics:

- An ACI recap
- The next technology
- Translating the features

ACI and the use case

Let's first revisit the use case we built upon in *Chapter 4*. Based on this use case, we determined that ACI was a technical fit for us:

> Your company provides an e-commerce solution to the market. Your solution is built on a number of different Microsoft Azure Technologies.
>
> Your product manager informs you that they have purchased licensed code to help them process billing statements. This code is an extension of the existing e-commerce platform and has a number of specific characteristics. The billing statement processing is not a continuous process. The processing happens when a customer requests their billing statements, which can happen randomly at any time of day. The processing is a sort of batch process with a peak load and needs to be available on demand. Deploying infrastructure to be continuously running requires considerable resource capacity, which makes it expensive, as the billing statements are generated continuously.
>
> Now, based on this use case, ACI would still be a very good fit. After all, the use case is still the same. However, let's skip forward a couple of years.
>
> The e-commerce solution is scaling rapidly and the usage of ACI for the billing statement container turned out to be a success. The supplier of the billing statement solution has provided a number of improvements over the past years, and their newest features contain a web frontend for their API. This web frontend can be rebranded to customer standards and runs in a container. The frontend container runs 24/7 and relies on the availability of the billing statement API. The billing statement solution can be run multi-tenant, but customers demand absolute separation from a security perspective. The solution will run single-tenant, and each customer is provided with their own instances of the solution and their own domain name. There are currently eight customers running the solution.

That is quite the addition! Instead of just using ACI for batch processing, we now have to run an additional container for the web frontend that runs 24/7. This is not exactly what ACI is billed for, and we will run into a number of problems if we continue down this road. It is time to rethink our container infrastructure.

The constraints of ACI

If we look at the use case and the way that ACI operates, we can see that there are some constraints to running the web frontend in ACI:

- The web frontend needs to be available 24/7.
- It is dependent on the billing statement API.

Can we technically run the web frontend on ACI? Yes, but it wouldn't be the most efficient.

ACI is built for batch processing, periodic jobs, and in general, for something that is not meant to run 24/7. Technically, it can run 24/7, but it is going to generate a significant bill at the end of the month. ACI is only charged based on usage; for every second you run it, you pay. That makes it very cost-efficient for batch processing or periodic jobs, as we don't have to pay for something that is running for a full month. However, with the addition of the use case where we introduce a web frontend, we do want it to be up and running 24/7. We don't want the customer to wait for the solution to "start" when they access it, but we also don't want to be charged for every second of the month.

That means we have to move to a different technology if we want the application to run as efficiently as possible.

This will be followed by an internal discussion within the company because running things as efficiently as possible does not always mean "cheaper." We are adding new features to our solution, which normally comes with an increase in infrastructure costs. In this case, it means running a technology for a full month and paying for that instead of periodic or incidental processing. In our experience, if you explain this well, everything is fine and everyone understands the increase in costs. If you don't explain this, it will result in questions afterward, as not everyone within your business will understand the technical implications. Additionally, **Azure Container Apps** (**ACA**) does not come with all the Ingress magic out of the box. We don't get a fully featured Ingress controller where we can manage things such as TLS and TLS termination. We need to make adjustments to our frontend container itself to make this work or deploy an nginx container to manage the routing and certificates for the web frontend.

We have plenty of motivation to move to a different technology. Let's take a look at what that might be.

Motivations for migrating to a different technology

To summarize, we can say that our motivation to research another technology is both financial and technical.

Financially, ACI will be running for a full month, as the web frontend requires that, which turns the ACI's efficient and low-cost serverless pricing model into the complete opposite – a very expensive technology.

Technically, ACI is not designed for running continuously. It is serverless by nature, and when unused, the solution shuts down. Only on the next "call" to the web frontend will it start again, which introduces unnecessary delays and a poorer customer experience.

In this section, we are only just scratching the surface of all the technical capabilities ACI can provide, and we are basing our motivation to migrate to another technology purely on the use cases of ACI and our fictional company. If we really dig deep and start looking at the actual technical features these products provide, we can find more *motivations* to move. That is why we are currently referring to this as a motivation to research. The mismatch between the platform use case and the fictional company is evident, and we have enough motivation to research another technology. Due to our previous research into containers, we know Azure App Service and Azure Functions are usually not the go-to technology.

The biggest question is, what is the next technology we can use and why?

The next technology

We have two choices. We can use ACA, or we can use **Azure Kubernetes Service** (**AKS**). You could argue that we could use Azure Web Apps as well – that is true, but it is not the best for running multiple containers. Keep in mind that both Azure Functions and Azure Web Apps have a feature that allows us to run containers, whereas the other three technologies (ACI, ACA, and AKS) are designed specifically to run containers.

If we look at the use case, we can deduct a number of technical requirements:

- The web frontend relies on the billing statement API.
- The solution (the frontend and API) is deployed per customer (single tenant).
- The solution runs 24/7 throughout the month.
- We need to service eight customers.
- We need a custom domain for each customer.
- The billing API and the web frontend have different life cycles.
- The billing API can generate a peak workload; therefore, we need scaling.

Technically, we can use both ACA and AKS, but we need to make a choice. That choice will be made based on size. As we extracted from the use case, we only need to service eight customers. This means we need to run 8 x 2 containers (one for the web frontend and one for the API). If we were to run this on AKS, we would introduce significant overhead, costs, and complexity for only 16 containers in total. If we expected the number of customers to double within the next few months, then AKS might be of interest and worth the investment, but we would argue that even then, you probably want to wait a little longer for the business plans/goals to become clearer to ensure you are picking the right technology.

On the other hand, ACA provides isolation out of the box and has the features to configure Ingress out of the box, allowing us to connect both the frontend and API to each other without the complexity of AKS. From a financial and learning perspective, this seems the better option *for now*; at some point, we have to migrate to AKS for efficiency reasons. As a business keeps growing and expanding, the overhead of a service such as AKS is no longer of concern.

We continue to repeat that it is only natural to move to the next platform every now and then; in fact, we have to. It is unacceptable for an infrastructure to be the limiting factor. Just as developers add new features to a solution through code, we will add new features to the infrastructure. It is never a one-time task and never a *fire-and-forget* exercise.

Another important reason to choose ACA over AKS, for now, is the learning curve. By using the other technologies to run containers on Azure, and perhaps all other Azure technologies we run for other parts of our infrastructure, we have become accustomed to tools such as the Azure CLI, Azure PowerShell, ARM templates, or Bicep. We have yet to discover the world of YAML, another language and standard you need to learn. When we start using AKS, we are not only tasking ourselves with understanding a fairly complex platform; we also need a thorough understanding of YAML, Helm, and how they can be applied to manage our AKS cluster. If you have been building Docker containers yourself, you already have some experience with YAML, but normally, these files are limited in size. When we talk about configuration management on AKS, YAML files can easily increase to hundreds of lines of code, and with all the different moving parts AKS provides, this can become complex. To a degree, the same goes for ACA; we can use YAML for more complex deployments, but for most scenarios, the Azure CLI will still suffice. What we are trying to say is, yes, you can decide to skip ACA if you envisage it not being of use for much longer, whether that is because the number of customers increases or you simply require more features. We do want to stress that you cannot underestimate this. Running AKS in production and at scale is different from "trying to see whether your container runs." There is a lot of planning and learning involved. Don't get us wrong – we think it is amazing, but we often see customers underestimate the leap to AKS.

However, with ACA, you are already learning about AKS without realizing it. If it runs on ACA, it runs on AKS. The management and complexity of Kubernetes are just abstracted away. This means that if your developers start embracing more modern event-driven architecture and microservices, ACA will be a good fit for a long time, at least until the business is mature and big enough to run AKS. This will also simplify the decision to ever move to AKS. A part of that decision will be technical, but sometimes, the decision will simply be made because *we have outgrown ACA, our business is doing well, and we want to invest in AKS.*

For now, ACA is the way to go in this scenario. But how does it compare to ACI? How different are they?

Translating features from ACI to ACA

Our next challenge is to translate the ACI configuration to ACA to make sure we can run our solution without impediments.

ACI configuration for the original use case

For ACI, we had the following configuration set up, as we discussed in *Chapter 4*:

- A container group per customer
- A billing statement API per container group
- A public endpoint for the customer to call and request billing statements

As the use case becomes extended over time and new requirements are added, our ACA configuration may look similar, but it is, in fact, different.

ACA configuration for an extended use case

Let's take a look at what configuration in ACA we need, according to our new use case:

- A container app environment per customer, as this will be the "secure boundary" our customers require
- Two container apps per environment, one for the web frontend and one for the API
- Ingress configuration per environment (customer)
- A custom domain for each customer

This is where the general concepts of both technologies can become somewhat complicated. It is pretty clear that instead of deploying to a container group in ACA, we are now deploying our solution to a container environment in ACA. That may look like a simple change of scenery for your container, but these two concepts are fundamentally different.

Container groups in ACI are not defined as a "secure boundary." In fact, they are a collection of containers that are scheduled at the same time, on the same host. Much like a Pod in AKS, the containers in a container group share the same life cycle.

On the other hand, ACA has individual life cycles, and each container app can contain one or multiple containers. A better way to compare the grouping between ACA and ACI is by saying, "Azure container groups translate to ACA," and the container app environment is an additional security boundary "around" the container apps per customer.

If we look at the use case and we need the web frontend and the API to communicate with each other but also scale independently, we will need two container groups. That brings additional complexity, as we now have to introduce Azure networking to enable communication between the container groups. Luckily, if we deploy them to the same virtual network, that communication is facilitated automatically. But now, we have deployed our container groups into a virtual network and the standard behavior with ACI is, if you integrate with a virtual network, there are no public endpoints anymore. Obviously, that would be a problem for our web frontend. To overcome this, we can deploy an application gateway and connect it to the virtual network to provide that access as shown in the following diagram:

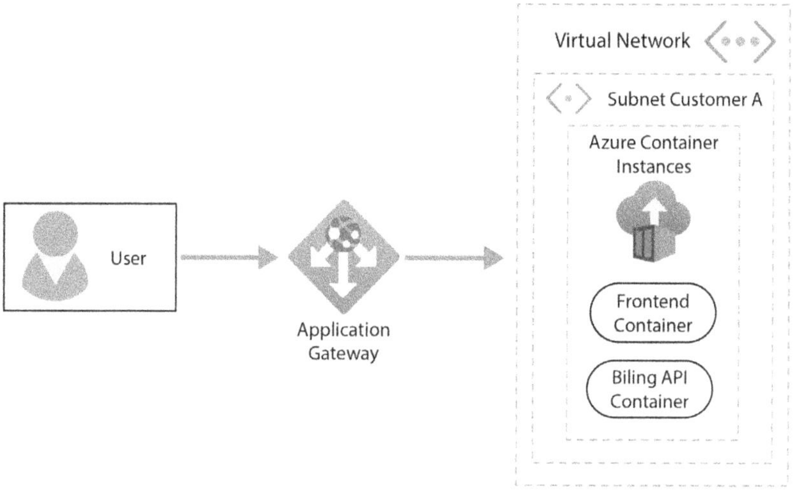
Virtual Network
Subnet Customer A
Azure Container Instances
User
Application Gateway
Frontend Container
Biling API Container

Figure 13.1 - ACI running in a vnet with application gateway

ACI also does not have the concept of *Ingress*. When deploying a container that is available externally, the address is similar to `frontend.eastus.azurecontainer.io`.

That doesn't look very friendly to a customer. Of course, we can do some magic with CNAMES, but the important thing is, the custom domains and certificates are not being handled in ACI. If you require such a thing, you need to build this into your container or use external services, such as Application Gateway, to introduce those features. Technically, it is possible but not very efficient.

You can see where this is going. ACI was a great technical fit for the initial use case, but now that we have additional requirements, ACI requires a lot of other configurations in Azure itself (virtual networking, application gateways, and so on). If we continue down this road, our ACI configuration might turn out to be more complex than a basic Kubernetes configuration. Not only will it be more complex, but it will also require more effort to manage the infrastructure, as we have more "moving parts" than before. Purely based on this, we can already make the decision to move on. And we did, as ACA provides the features that we currently need.

Let's summarize those requirements and see how they fit into either service:

Technical requirement	**Azure Container Apps**	**Azure Container Instances**
A separate life cycle for each container	The web frontend and the billing API will be deployed into a separate ACA.	To introduce separate life cycles, multiple container groups are required. Configuring the routing between those groups adds additional complexity and requires networking configuration in Azure.
Scaling	Azure Container Apps scales like Kubernetes does, so we can even configure **Kubernetes Event Drive Autoscaling (KEDA)** to automate the process.	Scaling doesn't exist in the concept as we normally know it. With ACI, scaling means that multiple distinctive instances of a container will be deployed.
A custom domain	Supported	Not supported
Ingress/HTTPS access	Supported	Not supported – requires your own nginx container if you want similar features
A secure boundary and isolation per customer	Azure App Environments are the secure boundary.	Not supported – although container groups can be configured to be secure, it is not the general concept of container groups.

Table 13.1 – A summary of requirements and technology best fits

As we can see, both technologies would technically be a fit, but whereas ACI requires a lot of additional configuration and Azure technologies, ACA can fulfill these features out of the box.

As for migrating, there is not really much to "migrate;" we simply did not have the features and configurations in ACI in place. In this case, migrating means deploying the container image itself, somewhere else. The great thing here is that we are not talking about data migrations or complicated swapping of DNS records. Customers just did not use those features before. If anything, we are deploying the next version of our solution to a new platform and then moving customers over. The "adding new features" part is particularly important here. Getting a customer on board for a migration to a different platform is much easier if we can promise them all the new stuff they are getting, such as a new web frontend.

Summary

In this chapter, we looked at how a use case can change over time. As we keep repeating, businesses change and requirements change. We have learned that migrating doesn't always mean that we have to rebuild exactly what we had but on a different technology. If we make the decision in time and we continuously try to improve and evolve our platform, we are not even speaking of a migration. We simply leverage features on the new technology, instead of going above and beyond and building a complex configuration on the current technology we are using.

We also learned that ACI is a very good fit initially, but when requirements change and customers become more demanding, we *have* to move to another technology, or we will end up with that complex configuration. Not only will that increase management effort and costs, but it will also make future migration much more complicated. To summarize, keep evaluating, make decisions, and make them on time.

In the next chapter, we are going beyond ACA – we will see whether we can take this use case even further, into AKS!

Summary

14

Beyond Azure Container Apps

We have successfully deployed ACA based on our extended use case. We have seen that it really fits the requirements of our use case. But here we go again: requirements change, customers change, and in this case, the number of customers is changing.

ACA was introduced as the *stepping stone* to AKS. In this chapter, you will learn that this is exactly what we used it for. However, the term *stepping stone* does not mean it is just a short-term solution. Sometimes, your solution runs just fine on ACA; there aren't many changes and there is no reason to migrate to something else. New features are still being added regularly as ACA is quite new. Perhaps, in 2 years' time, we will run even bigger workloads on ACA. But, it is a platform-as-a-service offering, which means your workloads and configuration are required to be standardized and you need to operate within the constraints of the platform. Sometimes, as you get more ambitious, the platform will become a limiting factor, especially in the world of container technologies. Sometimes, it is time to move on and explore new technologies, and that is why we are going beyond ACA.

In this chapter, we're going to go over the following topics:

- Azure Container Instances recap
- The next technology
- Translating the features

ACA and its use case

Let us quickly revisit the use case from the previous chapter. Based on the use case, we made a compelling argument as to why we needed to move on. But times have changed, and we might have to revisit the choice for ACA as well!

> Your company provides an e-commerce solution to the market. Your solution is built on several different Microsoft Azure technologies.
>
> Your product manager informs you that they have purchased a licensed code to help them process billing statements. This code is an extension of the existing e-commerce platform and has a number of specific characteristics. The billing statement processing is not a continuous process. The processing happens when a customer requests their billing statements, which can happen at any random time of day. The processing is sort of a batch process with a peak load and needs to be available on demand. Deploying infrastructure to be continuously running would require considerable resource capacity, which makes it expensive as the billing statements are not being generated continuously.
>
> The e-commerce solution is scaling rapidly, and the scalability and deployment, and costs when running ACA are becoming an issue. In the past 2 years, the organization has grown from 8 customers when starting out with ACA to 25. Considering the growth and the total number of containers (25 * 2), the organization is concerned about the number of resources, costs, and scalability. Additionally, the organization started developing additional containers for new services of their solution.

That is only a small addition to the use case but with major technical implications, and we might even be able to run on ACA for quite some time. Based on the numbers themselves, we don't really need to change technologies, but there is one very important addition: "Additionally, the organization started developing additional containers for new services of their solution."

Based on that quote, we cannot determine how big the number of containers will eventually be. As we have learned from the use case, this is a single-tenant configuration. Any new "service" might potentially be deployed to each customer.

The ideal scenario here is to see whether there are "shared" backend services, which is the case in most scenarios. However, we cannot assume this, and the numbers are already a compelling business case for a new technology such as AKS.

We also don't know what the requirements are for those new services. That means we need to look closely at the platform we are using. How flexible is ACA and can we trust that this will still be the technology for the years to come and still be suitable to our requirements? Perhaps not.

Based on the addition to our use case, we can determine new constraints that surface when using ACA at scale.

Constraints of ACA

We are not really hitting any major technical constraints with the current technology. Technically we cannot leverage more than two vCPUs in a single container app. But then again, we are not deploying multiple containers into a single container app. And let us just assume that our containers currently do not need more than two vCPUs.

We are also not running any privileged containers (which is not possible on ACA). It is also unlikely we ever will. Now that the organization decided on focusing future development on container technologies, we can guide them on the best practices – for instance, not running privileged containers.

The biggest constraint here is that we are already running 16 container apps spread over 8 container environments. Simply adding customers according to the use case means we will have at least 50 container apps spread over 25 environments. From an Ops perspective, this will become quite challenging. How will we manage all that? If we are doing infrastructure as code and using source control, deployments are easy. However, because we are currently creating customer-specific (single-tenant) configurations, that is a lot of code to keep track of, even in source control. Additionally, we have no idea what our developers will be building in the near future now that a business decision has been made to build new services on top of container technologies. The only thing we know is that they will run on containers.

Let's summarize:

- The number of containers is growing, which is cause for concern from an Ops perspective.
- The organization is developing new services based on container technologies; we don't know what the future growth will be but it is safe to assume it will be significant.
- We are not hitting any technical limitations of ACA just yet.

It is now time to look at our motivation for moving to a new technology.

Motivations for migrating to a different technology

The primary motivation is purely operational. We see customer demand is changing but we also see the internal business is changing. After all, the decision has been made to build more things on container technologies. With the current expected growth (25 customers), we need to run at least 50 containers (remember, both the billing API and frontend). If our developers are writing additional services and we run them in containers and they are single tenants, we will end up with 75 containers in total if just a single feature is added.

As an architect, business decision-maker, or lead developer, it is your task to minimize this amount. But really, if they are small and efficient services, are 75 containers really a problem? On ACA, this might become a daunting task due to the extra Azure resources needed, but for services like AKS, which were built to run a large number of containers, it is only normal.

However, it is a very good practice to evaluate your software architecture. In these use cases, the organization decided that single tenancy is the way to go, mainly from a security perspective. That doesn't mean all services to be built are containing sensitive information; maybe they do simple processing and can be run as a *shared backend*. This minimizes the number of containers, overhead, and management. Either way, there is no right or wrong way here, as long as you make an informed decision.

What is very important is that, unlike previous chapters and examples, this decision is not made from *higher up* in the business. This decision is likely led by your Ops team (or by you if you are the Ops team). That means the organization needs to be convinced of the upcoming *migration*. Usually, you would do this in the form of a small business case where you highlight how much time is currently spent on infrastructure management and how much the business will benefit from moving to a new technology. The biggest factor to highlight here is the risks. If we were to continue to run ACA, there are a lot of moving parts to maintain and plenty of infrastructure as code to maintain, as we are deploying customer-specific (single-tenant) configurations. It is important to note that there will not necessarily be fewer things to manage with a new technology, but there are definitely better technologies that allow for managing customization at scale.

Let us take a look at the next technology.

The next technology

We are just throwing it in there: the next technology we migrate to might very well end up being AKS! Yes, we can run everything on Azure Container Instances and ACA and it will run, but it won't be the most efficient.

Let us take a look at the new facts:

- The web frontend relies on the billing statement API.
- The solution (frontend and API) is deployed per customer (single tenant).
- The solution runs 24/7 throughout the month.
- We need to service 25 customers (new requirement).
- We need a custom domain for each customer.
- The billing API and the web frontend have different life cycles.
- The billing API can generate a peak workload, therefore we need scaling.
- New additions to the platform will be containerized (new requirement).

Even though there are only two new requirements, they make all the difference. As mentioned before, 25 customers mean at least 50 containers. New services (if deployed for each customer) cause exponential growth. We need a new technology that allows us to quickly add those additions (or a technology that supports a technology that can do this).

In the previous chapter, we wrote (quote), "*...because at some point we will have to migrate to Azure Kubernetes Service for efficiency reasons....*" And, we have just come to that point.

When we wrote the previous chapter, we highlighted that it was possible to skip the ACA step and go straight into Kubernetes. The biggest constraint was knowledge. Hopefully, you have become accustomed to building, running, and debugging containers over time and have a good understanding of ACA in general. That knowledge will make your move to AKS much easier. But don't underestimate the move; we are still talking about a significant change in infrastructure.

The biggest challenge is to not be blinded by all that AKS has to offer. Even though it is a lot, we don't want to implement everything we will ever need from day one. The best thing to do is stick to the use case and see what we need, standardize it, deploy it, and then look for future improvements. The *what* and *how* of those improvements is something we will look into in the next chapter: *Azure Kubernetes Service – The Next Steps.*

For now, we want to keep it simple. We want to merely copy what we have on ACA, follow best practices, and deploy an AKS environment that is up to the task.

AKS configuration

To deploy the cluster, we can follow the steps we discussed in *Chapter 6*. On top of that, we need to configure an Ingress, as discussed in *Chapter 8*.

Then, for security requirements, we need two additional configurations:

- Azure AD Authentication for AKS:

  Chapter 11 in the *Enabling Azure AD authentication and RBAC* section
- Network policies enabled for the cluster: `https://learn.microsoft.com/en-us/azure/aks/use-network-policies`

The Azure AD authentication is something we will use for administrators to log in to the cluster. This way, we can leverage technologies such as Privileged Identity Management to control access. The Azure AD integration provides more features such as role-based access control but, for now, we will focus on authentication to the cluster.

The network policies are going to be used to *recreate* the concept of container app environments. As we are going to use AKS, there is a learning curve to overcome. We have discussed this several times throughout the book but there are some basics we need to understand before transitioning from ACA to AKS.

Basic knowledge requirements

As we keep stressing, there is quite a learning curve for using AKS. A good understanding of the platform and its capabilities is key to successfully running your environment on AKS. It is very difficult to learn everything there is to know, and you could argue that there is nobody on earth that knows everything there is to know about Kubernetes. But, there are some knowledge requirements for getting started.

First of all, you need to be familiar with the YAML language, as we have seen in previous chapters. It isn't too difficult, and it is very readable, but it is important to understand how it works and how it translates to a running configuration in AKS.

Deploying the environment itself usually isn't the problem; if you follow the chapters we have written on AKS, we are positive you can get this up and running by simply replacing some code in the examples. But what about once it is running? How do you support it? This is usually when things will get troublesome: "*how do I debug something that is running, or not running?*"

It is required for you or your Ops team to familiarize yourself with *kubectl*. We have shown examples before, but this is the tool for your day-to-day troubleshooting of the cluster. *kubectl* comes with many features and options, but for daily operations, you only need a few. Therefore, we recommend you read up on *kubectl* in general and use the cheat sheet for daily reference: `https://kubernetes.io/docs/reference/kubectl/cheatsheet/`.

Last but not least, it is imperative you have a good understanding of the following concepts:

- Namespaces
- Deployments
- Pods
- Services
- Ingress

If you are unfamiliar with these, please read the chapters mentioned in the previous *AKS configuration* section to help you get started. Understanding these concepts, how they interact, and how you can leverage them to deploy your customer solution is key; the components make up the very core of what you are going to deploy and, from a technology perspective, can be mapped to ACA configuration. Let us see what that looks like.

Translating the features from ACA to AKS

Now that we have decided on AKS and looked at what we need to deploy, we need to understand how our ACA configuration maps to AKS.

ACA configuration for the original use case

For ACA, we had the following configuration set up, as we discussed in *Chapter 13, Azure Container Instances – The Next Steps*:

- One container app environment per customer, as this will be the "secure boundary" our customers require
- Two container apps per environment, one for the web frontend and one for the API

- Ingress configuration per environment (customer)
- A custom domain for each customer

The use case was extended; we need to make sure we can now run these services on AKS and continue daily operations. We don't really have new technical requirements but we need to translate exactly what we have here to an Azure Kubernetes configuration.

Azure Kubernetes configuration for an extended use case

Let's take a look at the configuration we need in ACA according to our new use case:

- A namespace per customer with network policies configured to recreate the "secure boundary" in ACA
- Two containers per namespace, one for the web frontend and one for the API
- Ingress resource per namespace (customer)
- Custom domain per customer

This does not look much different from our previous use case. We can rebuild the same concept/architecture in AKS. However, from deep within the technology, it is quite different. Instead of using app environments, we are now using namespaces. By default, a namespace is not secure, meaning that from a networking perspective, resources in different namespaces can communicate with each other. To fulfill the requirement of separation between customers, we recommend using network policies as referred to in the previous paragraph. This will allow us to limit traffic inside the namespace and only between the customer Pods.

Then we have the concept of customer Pods. These translate to the container apps themselves. As with container apps, a Pod can contain one or multiple containers. In this scenario, we are configuring one container per Pod, as we currently do not have a requirement where we need to run multiple containers side by side in a single Pod. We are deploying Pods using a *deployment*, which pretty much provides the *self-healing* capabilities of Kubernetes such as health probes, replicas, and much more.

> **Important note**
>
> In previous chapters, we discussed Dapr and that it uses a "sidecar" pattern. Normally, you could translate that to running a second container in the same Pod, which is often done when using a reverse proxy for things like nginx or OAuth2 configurations. If you have been using Dapr in ACA, you can install Dapr on your AKS cluster and add the annotations for Dapr in the deployment of your container.

We also have to make sure our customers can access their solutions on the cluster. For that, we will use an Ingress controller and Ingress resources to configure the custom domain. For certificates, we need to make sure the certificate exists as a secret in the customer namespace. For routing, that means an Ingress object per namespace (customer).

You could say, "That is all there is to it," but we understand there is more work to be done as opposed to running ACA, as we have to configure and manage more *moving parts*. Please note that all configuration is done through YAML, in code. Once you understand YAML, it is very scalable across customers. The downside is that it takes some time to get used to and set it up initially. The benefit is that you will be able to easily make changes and control your environment once you have everything ready to go.

Let's summarize those requirements and see how they fit in either service:

Technical requirement	Azure Kubernetes Service	Azure Container Apps
A separate life cycle for each container	The web frontend and the billing API will be deployed into separate Pods.	The web frontend and the billing API will be deployed into separate ACA instances.
Scaling	Pods can scale when deployed through a "deployment," and these support the configuration of so-called "replicas." If required, KEDA can be installed on the cluster. Additionally, nodes can also auto-scale, which is a benefit over ACA.	Azure Container Apps scales like Kubernetes does, and we can even configure KEDA to automate the process.
Custom domain	Supported through Ingress controller and resources.	Supported.
Ingress/HTTPS access	Supported through Ingress controller and resources.	Supported.
Secure boundary and isolation per customer	Supported through a combination of namespaces and network policies per customer.	Azure App Environments are the secure boundary.

Table 14.1 – Summary of requirements and technology best fit

We have determined before that ACA will still fit the use case from a technical perspective. After all, everything is supported. The change in technology is primarily because AKS is simply more efficient when it comes to deployment at scale. As we can see in the preceding table, all previously stated requirements can be met by applying the correct configuration in AKS. At scale, AKS provides more benefits but with a higher learning curve.

Summary

In this chapter, we have learned that ACA and the configuration we used for app environments and container apps themselves can be translated into an AKS configuration relatively easily. The challenge lies in understanding the platform and the tasks that need to be performed. This is quite different from the *out-of-the-box* features we have seen in ACA. However, it is necessary to move on to fulfill the requirements of our use case. We have learned what the technical implications are when moving from ACA to AKS.

That is not to say things cannot be simpler, but for things to be simpler, we first need to understand the platform and how our solution runs on it. In the next chapter, we will look at simplifying things from a deployment perspective. Do we really need so many YAML files per customer? Do we really need to deploy certificates to each namespace manually? Probably not, and that is what the next chapter is about: being more efficient and leveraging the features AKS provides to make life easier.

15
Azure Kubernetes Service – The Next Steps

We ended the previous chapter by stating that we can do things more efficiently on Azure Kubernetes Service and that is true! Even though we have stressed the complexity of Kubernetes throughout this book, that complexity also gives us a lot of freedom.

Many companies and open source initiatives have invested time and money into building solutions, add-ons, and services on top of Kubernetes. A lot of them have even been adopted by the Cloud Native Computing Foundation (`https://cncf.io`), indicating that they are not only promising but also actively maintained.

But it's not only third-party solutions that add value to Kubernetes. In fact, if you look at the Microsoft offering, **Azure Kubernetes Service** (**AKS**), they have added a lot of plugins to integrate and interact with other Azure technologies, making life even easier. And that is the investment Microsoft has made over the past couple of years: making the managing and maintaining of Kubernetes clusters easier. The complexity of Kubernetes is no longer a concern when it comes to day-to-day operations and maintenance. The complexity is in understanding how different Kubernetes components operate because in the end, if you are building a solution that runs on top of Kubernetes, it is vital to understand how that platform *responds* to your solution. It is not unlike writing a traditional client/server application for Windows in the past. We still needed to understand the operating system before we could build an app that runs stable and efficiently.

Our next step is to discover more about AKS and how it can help us bring our solution to a higher level of maturity. For that, we need to know how to write scalable deployments and minimize customer-specific configuration.

Therefore, in this chapter, we're going to go over the following topics:

- AKS and our use case
- Scalable deployments

- Automatic scaling
- Windows containers

AKS and our use case

Overall, our use case requires us to use AKS because we need to comply with the following requirements:

- The web frontend relies on the billing statement API.
- The solution (frontend and API) is deployed per customer (single tenant).
- The solution runs 24/7 throughout the month.
- (New) We need to service 25 customers.
- We need a custom domain per customer.
- The billing API and the web frontend have different life cycles.
- The billing API can generate a peak workload, therefore we need scaling.
- (New) New additions to the platform will be containerized.

To implement the infrastructure to fulfill these requirements, we recommend an AKS cluster and the use of namespaces and network policies to create a secure boundary. Additionally, we will use an Ingress controller and Ingress resource to allow the customer to configure the custom domain and provide a certificate for secure traffic.

For security reasons, we also recommended the use of Azure **Active Directory** (**AD**) integration for administrative operations on the cluster itself. So far, so good, and technically that setup is enough to get everything up and running and provide our services to customers.

However, we have also identified some complexities. As we are running a single-tenant solution, we are now tasked with creating different configuration files (YAML) per customer. This wasn't much different in Azure Container Apps; every customer was a unique deployment. Of course, we can do smart things in code to achieve this and minimize the lines of code; but in the end, it is still a configuration per customer. We want to simplify that to ensure changes are made to each and every customer and not missed. This can be achieved by using a tool called Helm, which packages up your YAML files into template-driven deployment packages.

Additionally, we are tasked with certificate management. Whether we provide a unique URL based on customers' wishes or we create a URL such as `customerX.cloudadventures.com`, we still need to make sure we store a certificate in the namespace of the customer for the Ingress object to use.

In the current use case, we are also leveraging Azure AD for authentication to the cluster but we have not implemented role-based access control specific to the customer namespace. The big question is: does everyone always need access to all namespaces or can we introduce some granularity there?

Then, we still need to look into some things we call *hygiene factors*. One thing we haven't talked about yet in this use case is monitoring and troubleshooting the AKS cluster itself and the workloads that run on it. We also haven't looked at automatically scaling our environment when customer demand increases. With the concept of the billing API, which can generate a peak workload, that is also something we definitely need to look into.

These are some improvements we can implement to take our AKS cluster to the next level. To summarize, we want the following:

- To write our *infrastructure as code* or YAML more efficiently
- Role-based access control for the namespaces
- To simplify certificate management
- To implement proper monitoring
- To automatically scale the cluster/nodes

Luckily, there is a technology to help us implement all of this and we have already touched up on most of those technologies and explained how to use them in *Chapter 11*. In that chapter, we discussed the use of role-based access control in combination with Azure AD, using Azure Key Vault for secrets management and certificates, and how to enable and configure Azure Container Insights.

Of course, in *Chapter 11*, we built the cluster from scratch, but now that we have "moved on" from different technologies, you can simply enable these features whenever you wish to do so.

That means we still have some new things to explore. Specifically, we are going to look at making our deployments scalable and better leveraging infrastructure as code. Additionally, we will investigate how we can automatically scale our cluster to keep up with customer demands.

Let's first start by looking at making our deployments more efficient. We will do that with the use of the Helm package manager for Kubernetes.

Scalable deployments

Scalable deployments are something we hear about quite often; it might even be considered a buzzword. But it is especially very important with Kubernetes. Writing infrastructure as code to deploy solutions at scale is something we could classify as a requirement for using Kubernetes. Throughout this book, we have started with small deployments; over the course of a couple of chapters, we went from one customer to eight, and now we are preparing our infrastructure to deploy an infinite amount of customer solutions.

Doing all this the traditional way using the YAML language can become very chaotic. Normally, in YAML, we would write a file for a customer to deploy the frontend as we did in *Chapter 11* (`https://github.com/PacktPublishing/Azure-Containers-Explained/blob/main/CH11/frontend.yaml`).

If we inspect this file, we can see some hardcoded values that ideally we would parameterize. Let's look at the following example for an Ingress resource:

```
apiVersion: networking.k8s.io/v1
kind: Ingress
metadata:
  name: frontend
spec:
  ingressClassName: internal-ingress
  rules:
    - host: cloudadventures.com
      http:
        paths:
          - pathType: Prefix
            backend:
              service:
                name: frontend
                port:
                  number: 80
            path: /
```

Of course, we want to parameterize as much as possible, but if we look at the preceding example and take into account that we want to have a unique deployment for each customer, then we need to at least parameterize the hostname. Going a bit deeper and inspecting the entire `frontend.yaml` file, we can also determine that it is probably a good idea to name the deployment after the customer as currently, it is referred to as `frontend`. Now, the latter is not a hard requirement; if we deploy to a separate namespace, it will run just fine, but for readability and day-to-day operational management on the cluster itself, we would very much appreciate names that we can use to identify the customer itself. Sure, we can use labels for that, but if we are in the process of parameterizing our deployments, we might as well go all the way.

The biggest question right now is: how do we do that? In the world of Kubernetes, there is a standard for that, the Helm package manager (`https://helm.sh`). Let's take a look at converting our `frontend.yaml` file into a simple Helm chart.

Before you can create a Helm chart, you need Helm installed. You can do this in many ways, depending on your operating system. You can check out the current installation methods at `https://helm.sh/docs/intro/install/`. For this, we are going to use Azure Cloud Shell, which already has Helm installed.

Creating a Helm chart

Helm has a built-in option to allow us to create a starter chart. We are going to use this as a base. In Cloud Shell, run the following command:

```
helm create frontend
```

This will create a folder called `frontend` (the chart name) and inside this folder are all the files and directories that make up a Helm chart:

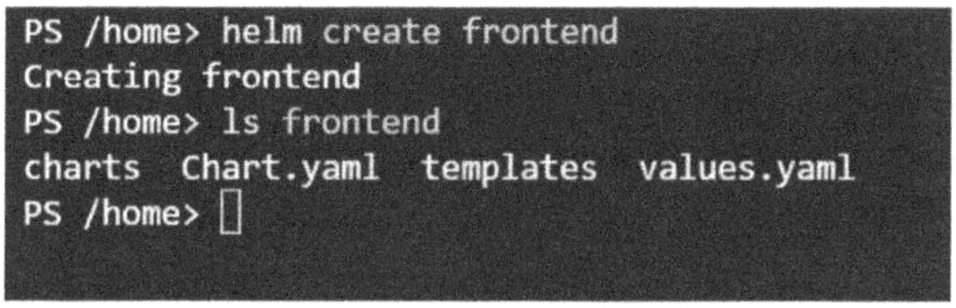
PS /home> helm create frontend
Creating frontend
PS /home> ls frontend
charts Chart.yaml templates values.yaml
PS /home>

Figure 15.1 – Output from helm create and ls commands

The folder structure looks something like this:

```
frontend/
  Chart.yaml
  values.yaml
  charts/
  templates/
```

Let's look at each folder and file and see what it is used for:

- The `Chart.yaml` file contains a description of the Helm chart, the chart version number, and your application version number. It is recommended you update this file with your details. For this demo, we will keep the defaults.
- Next up is `values.yaml`. This is a very important file. It contains all the default values for your chart. These values can be overwritten by creating other `values.yaml` files, say one per customer, or by passing in the values at the deployment time.
- The `charts` directory may contain other charts, better known as subcharts, in the Helm world. Subcharts are out of scope of this demo, but something we would advise you to look into.
- Finally, we have the `templates` folder. This folder is where the templates live. When I say templates, they are basically the same YAML files we are familiar with but with placeholders that are overwritten using the `values` file.

Inside the `templates` folder, you will see a lot of files and another directory. This test directory is where you can store template files to test your deployment. They are out of the scope of this demo, so go ahead and delete the directory using `rm -rf tests`.

You will also notice some other files we don't need. We know that our current YAML comprises three Kubernetes objects: a Deployment, a Service, and an Ingress object. So, it is safe to delete all other files in the `templates` folder apart from `deployment.yaml`, `service.yaml`, `ingress.yaml`, and `_helpers.tpl`. You can use the following to do that: `rm ./hpa.yaml ./NOTES.txt ./serviceaccount.yaml`.

With that, you have your Helm chart ready for a test deployment, but before you can deploy the chart, you need to create a new `values` file and supply the information to overwrite some parameters. Go ahead and create a new YAML file called `customer1.yaml` in the root of the frontend folder alongside the current `values` file.

Here, we need to provide any values that will be different for each customer. For this example, use the following:

```
nameOverride: customer1
fullnameOverride: customer1

image:
  repository: whaakman/container-demos
  # Overrides the image tag whose default is the chart
appVersion.
  tag: cloudadventuresshopv1

ingress:
  enabled: true
  className: internal-ingress
  hosts:
    - host: customer1.cloudadventures.com
      paths:
        - path: /
          pathType: Prefix
```

By supplying the values, we are overwriting the default values. So, the name will now be `customer1`, the image repository is pointing to Docker, and we have specified the tag separately to make it easier to update for the next release. We then enable Ingress by using the `true` switch and pass in the customer-specific hostname.

Deploying the Helm chart

It's now time to test to see whether we have everything right. Make sure you are connected to an AKS cluster as the command we will use will run against the Kubernetes API. We can use Helm's dry run for this. In Cloud Shell, navigate to the root of the `frontend` folder and run the following:

```
helm upgrade --install customer1-frontend ./ --values ./
customer1.yaml --dry-run
```

Once that runs, you will see some YAML outputted in the terminal, as in *Figure 15.2*.

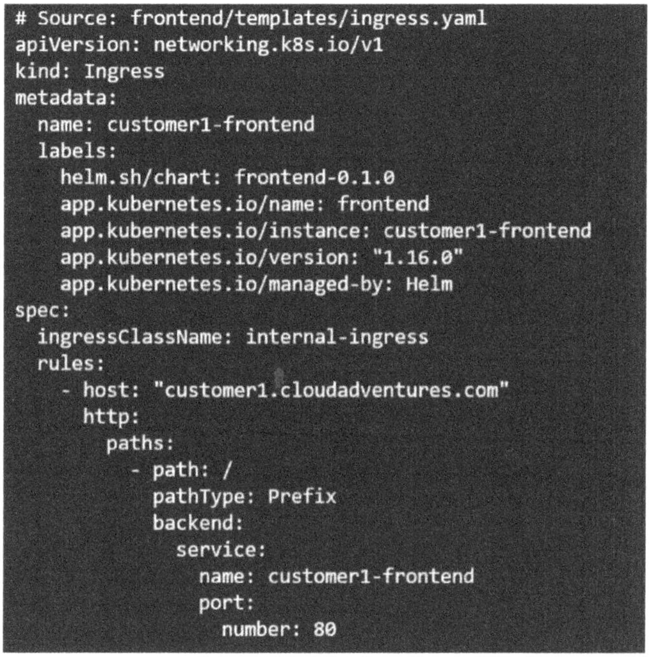

```
# Source: frontend/templates/ingress.yaml
apiVersion: networking.k8s.io/v1
kind: Ingress
metadata:
  name: customer1-frontend
  labels:
    helm.sh/chart: frontend-0.1.0
    app.kubernetes.io/name: frontend
    app.kubernetes.io/instance: customer1-frontend
    app.kubernetes.io/version: "1.16.0"
    app.kubernetes.io/managed-by: Helm
spec:
  ingressClassName: internal-ingress
  rules:
    - host: "customer1.cloudadventures.com"
      http:
        paths:
          - path: /
            pathType: Prefix
            backend:
              service:
                name: customer1-frontend
                port:
                  number: 80
```

Figure 15.2 – Output from the Helm dry run command

If we compare *Figure 15.2* with the original Ingress YAML, we will find it is basically the same apart from some added labels (which were added by Helm) and we now have `customer1` in the name and the hosts.

If we wanted to deploy `customer2`, we would need to create another YAML file called `customer2.yaml` and change `customer1` to `customer2`. Then, when we run the `helm upgrade` command again, we just use this new YAML file.

If we look at the newly created `ingress.yaml` file in the Helm chart, we will see it is now around 60 lines compared to the 17 from our previous examples. Now that is a lot of extra lines, but we did not have to write them and now we have the ability to easily reuse the chart for each new customer by just supplying a few lines of values.

This was a super simple example of creating a Helm chart to get you started. We highly recommend you read more about Helm and how to use it to its full potential. We have uploaded the frontend Helm chart to the Git repo for you to have a look at. You can find it at `https://github.com/PacktPublishing/Azure-Containers-Explained/tree/main/CH15/frontend`.

With the application now in a Helm chart making repeatable deployments easier with the usage of `value` files, it's time to look at automatically scaling the AKS cluster and the workloads running on it.

Automatic scaling

Hopefully, we have allocated the correct capacity to our AKS cluster to make sure daily operations run smoothly. However, customer demand can change and if our business is running well, we are also adding customers. Ideally, we automatically allocate resources to our cluster once these demands change.

Before we can enable technologies that allow for automatic scaling, it is imperative that we understand how much capacity is needed to initially run our resources. For example, if we are unaware of how much we actually need, we may be presented with an invoice at the end of the month that we didn't expect. In order to prevent that, we can set limitations to our autoscaling, but that also means that at some point, the environment will no longer scale and it will impact the customer experience. The more we know about the behavior and capacity requirements of our solution, the better we can configure autoscaling.

When we talk about scaling, we are mainly talking about horizontal scaling. This means we are adding an instance of a node to the existing node pool or we are adding another instance of a Pod to our deployment. This is opposed to vertical scaling, which means we are adding more capacity to a single instance. In general, vertical scaling involves restarting a Pod or, in the case of node pools, deploying a new node pool. For continuity purposes, we always want to focus on configuring horizontal autoscaling.

Scaling the nodes

In AKS, we have a node pool. This node pool runs on a virtual machine scale set with *X* number of instances (we recommend a minimum of three to ensure redundancy over the three Azure availability zones). These instances have a specific SKU that is fixed for that node pool. In the scenario where we add additional customers, at some point, we are going to need another instance of the node. If we use manual scaling, that means we have to run into performance issues or Pods unable to be scheduled before we take action (or are alerted by our monitoring that we are running at capacity). But what if we were to configure the cluster so that it automatically adds a node when such a scenario occurs? We can configure the cluster autoscaler by adding it to an existing cluster if we have one or we can also enable it at cluster creation.

To update the settings of our cluster, we use the `az aks update` command with the required parameters. To enable the cluster autoscaler, we run the following command:

```
az aks update `
  --resource-group AKSResourceGroup `
  --name AKSClusterName `
  --enable-cluster-autoscaler `
  --min-count 3 `
  --max-count 6
```

We pass the name of the resource group that our AKS cluster is in, then provide the cluster name and enable the cluster autoscaler.

Previously, we also discussed the importance of limits. We don't want to end up with an incredible number of nodes because we made a mistake somewhere. This is where we can set the minimum and maximum count of nodes.

> **Important note**
>
> When running multiple node pools, we need to define which node pool we want to enable or update the cluster autoscaler for. This is done through the `az aks nodepool update` command and as an addition, `--name` becomes `--clustername` and `--name` is used for `NodePoolName`.

The cluster autoscaler works "both ways." If demand is high, nodes are added, but if demand is decreasing, nodes are removed. Additionally, we can configure an autoscaler profile and add granularity to our scaling configuration, such as scan intervals and thresholds. But please be aware that the scaling profile applies to all node pools that use the cluster autoscaler.

Now that we know how to enable the automatic scaling of the nodes and can be sure that capacity is available when we need it, we can look at scaling the Pods.

We don't have to configure the scaling of nodes and Pods in this specific order, but it makes sense to ensure that when our Pods scale, the nodes will also scale to accommodate the new Pods.

Scaling Pods

As seen in the previous section, if we can scale nodes, surely we can scale Pods as well. For this, Kubernetes uses the **HorizontalPodAutoscaler** (**HPA**). To configure this we need to configure an `HorizontalPodAutoscaler` object. This object contains all the configuration, such as the metrics you require to trigger the automatic scaling. For instance, we can configure the minimum and maximum number of replicas (instances of the Pod) and that it should scale when a certain CPU or memory limit is reached.

When configuring an HPA object, we *link* it to an existing deployment by configuring `scaleTargetRef` and referencing our existing deployment.

Let's assume we have a deployment called `frontend-cloudadventures`:

```
apiVersion: autoscaling/v2
kind: HorizontalPodAutoscaler
metadata:
  name: frontend-cloudadventures
spec:
  scaleTargetRef:
    apiVersion: apps/v1
    kind: Deployment
    name: frontend-cloudadventures
  minReplicas: 1
  maxReplicas: 10
  metrics:
  - type: Resource
    resource:
      name: cpu
      target:
        type: Utilization
        averageUtilization: 50
```

In the preceding example, the `HorizontalPodAutoscaler` on Kubernetes will look at the `frontend-cloudadventures` deployment and scale to an additional instance (replica) once the CPU utilization is above 50%. As with the cluster autoscaler for nodes, this works the other way around as well; if the utilization drops below 50% for a Pod, an instance is removed. Additionally, we have configured a minimum of 1 replica and a maximum of 10 replicas.

By combining both the node autoscaler from the previous section and the HPA from this section, when the deployment scales to 10 replicas and the nodes do not have enough capacity to schedule a new instance of a Pod, a new node will be provisioned as the cluster autoscaler is triggered, resulting in a new node with the new Pod(s) running. HPA can do much, much more. If you are interested, we recommend reading up on the latest features and best practices here: `https://kubernetes.io/docs/tasks/run-application/horizontal-pod-autoscale/`.

Scaling Pods with the HorizontalPodAutoscaler specifically looks at the metrics. We can go above and beyond and think of all kinds of metrics, but what if we want to scale based on more operational metrics, such as messages in an Azure Service Bus queue or blobs in an Azure Storage account? That is exactly the use case for KEDA (`https://keda.sh/`). We have already discussed KEDA in this book before but it is good to mention that KEDA leverages the `HorizontalPodAutoscaler` for its scaling operations. KEDA could be seen as an event-driven addition to the HPA. That doesn't mean you have to use KEDA, but it is important to understand that if you run into limitations of the HPA, KEDA is a good technology to explore.

Now that we understand the two types of scaling in AKS, node and workload, let's take a look at Windows container workloads.

Windows containers on AKS

Something we have only briefly touched on throughout this book is Windows containers. The majority of containerized solutions are based on Linux and we don't expect that to change any time soon. However, there are big use cases for Windows containers and one of them surprisingly is cost-based. Multiple technologies we have discussed before support Windows containers but they are not as feature-rich when it comes to Kubernetes. In April 2019, Kubernetes announced support for Windows containers. By attaching a Windows machine to the Kubernetes cluster, this machine would become a node that could run Windows 2019 container images. Soon after, Microsoft adopted this support in AKS. Initially, AKS provided only limited support for Windows containers, but the use case was there. We can now run .NET Framework solutions on AKS provided we provision a Windows 2019 node pool and build our containers on Windows 2019 images.

Use cases

Let's take a deeper look at that use case. Even though Windows containers have matured significantly on AKS in the past couple of years, it is probably still not the operating system to start out with when writing a solution from scratch. But, that's also not the use case of Windows containers.

Throughout the market and customers, we see the growing usage of Windows containers as a result of migrating legacy applications away from traditional infrastructure. Plenty of software companies have been around for years and likely have some legacy software architecture running on the Windows platform. In that playing field, we see the biggest use case for **Internet Information Services** (**IIS**) based workloads. That is because a container is stateless by nature. We don't store information inside the container and we don't "remote desktop" into a container. Solutions running as an IIS workload are a good target for containerization.

Legacy solutions

The use case for legacy solutions is, as we mentioned, mainly for IIS-like workloads. But it is not limited to these workloads. Services for (batch) processing that would normally be installed on the Windows machine can also be containerized and run perfectly on AKS.

In the past 2 years, we have seen a growth in software companies moving to the cloud. Some have already set out a strategy to move to containerized solutions and are still being held back by that one piece of legacy software running on IIS. And some companies don't have that strategy yet but really want to move away from their traditional data center and to the cloud.

The advice is, and will remain for time to come: move to Linux containers when you can. But for a software company that has spent years investing in its current .NET Framework solution, that is usually not a short-term scenario. It can take months if not years depending on the solution.

Moving a part of the solution that can't run on Linux containers to a Windows container architecture is a temporary solution but can run for quite some time. It is more efficient than running separate virtual machines for your IIS deployments, and a container-based architecture scales quicker than traditional architectures.

As a result, companies no longer have to invest in virtual machines or hardware in a data center, which can be a significant cost saving. However, usually, this choice is made from a technology perspective and not a cost perspective.

Cost-based

Another use case for using Windows containers is based on costs. We have to admit, we (the authors) didn't see this one coming but have seen some scenarios where this decision was made purely based on costs.

Azure App Service is a very popular technology on Microsoft Azure that provides you with the ability to run your .NET Framework-based workloads. The runtime is supported out of the box and if we look at software companies that run IIS workloads, Azure App Service is usually the first thing they will research.

Some companies run hundreds of App Services across multiple App Service plans. While Azure App Service can be a really efficient solution and you can minimize a lot of overhead if you write your software according to the provided standards, not a lot of companies do that. It's not that they don't want to, but that they can't due to the technical debt of an older runtime and framework.

This is where things become interesting. In our day-to-day jobs, and based on our experience with software companies, we have seen significant cost savings for companies who ran hundreds if not thousands of App Services and containerized that solution and run them on Windows node pools in AKS. In some scenarios, we are talking about 20-30% of Azure costs, just because Windows node pools are now quite efficient when it comes to overhead.

Another big win for those companies that moved from App Services to AKS is the upgrade scenarios. While for App Services you had to push to multiple App Services (or they pulled the code from a repository), with AKS and containers, you will simply redeploy the YAML (or upgrade Helm charts) with a new version of that image. That image is then downloaded to the node(s) initially upon first deployment and can be used by other Pods. You can imagine this is a lot faster than upgrading all the App Services you have at the same time, in batches or even one by one.

Features of AKS and Windows containers

We have discussed that Windows containers actually make a compelling use case, but there are some things we need to share before you decide to give this a try.

First things first: the AKS control plane runs on Linux, therefore you always need at least one node pool of the *system* type running on Linux. Your secondary, or all, node pools after that can be Windows, but you need at least one Linux node pool.

It is safe to say that the majority of AKS features are supported for Windows node pools and Windows containers but there are some things to take into account.

Operating system version

You can only use Windows 2019 and Windows 2022 container images and node pools. We have to be aware that the operating system used for the container image must match the operating system of the node itself.

Image sizes

The Windows operating system requires more disk space and thus images are bigger than what you are used to when running Linux containers. When a container image is first used on a node, this image is pulled from the registry. For Windows containers, expect a longer startup time as opposed to Linux containers.

Updates

If you have administered Windows in the past, you might have the monthly update rotation down as a habit, and every month you are installing the latest Windows patches. On AKS, that works differently. We no longer use the traditional Windows update. Instead, AKS releases new node pool images (for both Linux and Windows) as soon as patches are available and you can upgrade your node pools by applying those images.

Networking

If you are already using AKS, we hope you went with our recommendations to use the **Container Networking Interface** (**CNI**) to integrate AKS with Azure Virtual Networks. The alternative to that is to use the default **Kubernetes Networking** (**Kubenet**). For Windows node pools, you have to use Azure CNI as Kubenet is not supported. Please be aware that the selection between Kubenet and CNI is made on cluster creation and cannot be changed afterward. If you already have an AKS cluster and are using Kubenet, this means you have to deploy a new cluster with CNI or redeploy the existing cluster.

Azure Hybrid Use Benefit

Windows node pools are essentially still Windows servers and you are still required to pay for the Windows Server license. In Azure, the license costs are automatically charged through virtual machine pricing. But what if you were going with the scenario of moving from on-premises virtual machines to Windows containers? Perhaps you already have Windows Server licenses that you want to migrate to the cloud. For this, Microsoft provides Azure Hybrid Use Benefit. If you have eligible licenses, you can reuse them in Microsoft Azure, and as a result, the license costs are deducted from the monthly virtual machine (scale set) costs. This also works for Windows node pools!

Third-party integrations for Kubernetes

As we have mentioned before, when it comes to third-party integrations and Kubernetes, the sky is the limit. There are many open source solutions out there that can add value to your Kubernetes cluster. However, many of these solutions were never developed with the intention to support Windows containers. It is important that when you are running a hybrid configuration where you run both Linux and Windows node pools to research these solutions before trying to apply them. Some simply do not support Windows.

The solution to that is to run a separate cluster just for Windows containers, but whether that is beneficial from a financial or technical perspective really depends on your specific use case.

The preceding features really are the most important things you need to consider when running Windows containers. In the past, there were plenty of constraints, but Microsoft has made significant investments into Windows containers over the past couple of years, resulting in a stable and trustworthy solution when it fits your use case.

Summary

In this chapter, we have learned what we can do to improve our basic AKS cluster used in the use case of *Chapter 14*. Some technologies were already discussed in earlier chapters (such as Azure AD integration, role-based access control, and integration with Azure Key Vault). However, we have introduced the basics of scaling nodes and Pods and how the cluster autoscaler and the HPA both operate on different levels but complement each other. We have learned how writing and managing Helm charts is a valuable skill as it helps with scaling your deployments without having to write extensive customized YAML files per customer.

We closed with a section on Windows containers, where we briefly touched upon some specifics we need to take into account when running Windows containers. We learned about use cases and have seen that there is definitely a use case for Windows containers out there.

In the next chapter, we are going to dive into more technologies related to containers, Azure, and Kubernetes that will help us increase the maturity of our platforms. Some of those technologies are still upcoming; some are underrated but very valuable. We will share our thoughts on these technologies with you and hope they will benefit you in the future.

16

What's Next – Diving into Future Trends and More

Over the course of this book, we have looked into different technologies provided by Microsoft Azure to run containerized solutions. We have discussed Azure App Service, Azure Functions, Azure Container Instances, Azure Container Apps, and Azure Kubernetes Service. These are the most popular and well-known technologies on Azure when it comes to containers. However, there are more. Not only that, but there are also more to come in the future as the container landscape is still rapidly developing and new initiatives are launched almost monthly.

In the cloud-native world, technologies are advancing even more rapidly than we were just getting used to in the world of the public cloud. If we want to keep up with what our developers are building, we need to stay on top of the new technologies being released and learn about their use cases. During the course of this chapter, we will look into new features that have recently been released, other technologies that might be of interest, and in general the trends that we see throughout the market.

By no means will this be a definitive list. After all, the landscape is developing faster than we could anticipate while writing the book. But we do feel we have selected the interesting technologies you might want to research now and in the future.

In this chapter, we're going to go over the following topics:

- Future trends
- The future of AKS
- Certification

Future trends

The container landscape is maturing. Where containers have been around for quite some time, only in the last couple of years has the growth in container usage been enormous. Instead of researching whether we can run something in a container, resulting in a *maybe in the near future*, it has now become a logical choice for the next version of a solution. Even though the technology of containers itself is very mature (Docker, containerd, Open Container Initiative), the orchestration of containers is something we are still struggling with from a maturity perspective.

As we have learned throughout this book, a lot of solutions end up in Kubernetes, which can be a very complex platform. We are not talking about the maturity of technology but about the maturity of organizations and teams using that technology. Often, they don't have the processes in place to successfully manage those technologies. The same could be said for IT in general as it is still relatively new to the world compared to other business areas. However, we see that the maturity is definitely increasing, and companies are embracing and enforcing standards and keep building more advanced and stable platforms on top of solutions such as Kubernetes. That brings us to future trends we see becoming a big part of businesses in the near future. Let's dive in!

Machine learning and AI

Machine Learning Ops (**MLOps**), now that is a buzzword we hear quite often these days. It's necessary, nonetheless. Machine learning models require continuous development and improvement; they are not unlike regular code for, let's say, an application. Normally, these models are deployed in the form of a web service where a client can connect to and consume said service.

So, just like a normal development life cycle, we want fast deployments to development and production environments. Additionally, Kubernetes in general is very suited to run machine learning models. With the concept of node pools that we learned about throughout this book, we can simply deploy a node pool with the computational power required to run the model, whatever that capacity requirement may be.

In the past, we would run these models on a compute cluster of virtual machines, which didn't have the features that, for example, an AKS cluster provides. Recently, we have seen a shift to machine learning models being deployed to Kubernetes, sometimes because it is just more efficient, especially from a scaling perspective, and sometimes because all other components are already running on Azure Kubernetes Service.

Either way, the way we perform ITOps for AKS and the capabilities it provides us to do ITOps really benefit machine learning. Together with the tight integration that Microsoft provides through various SDKs and Azure services between Azure Machine learning and AKS, we really expect the usage to grow significantly over the upcoming years.

Cloud-native development

Development is one thing and cloud-native development is another. We have experienced and seen multiple solutions being migrated to the public cloud. A *lift and shift*, if you will, sometimes paired with minor refactoring of the code base. Building true cloud-native solutions means taking the capabilities of the public cloud into account, adjusting your software architecture to those capabilities in order to build scalable and resilient solutions that thrive on the public cloud. What we have seen in recent years is that most of the growth in AKS usage has been containerizing existing solutions and hosting them on AKS. That doesn't automatically mean the solution is *fit* to run on AKS. What we have learned is that when businesses have migrated their solutions to AKS, they sometimes try to shoe-horn things into a container on AKS and have to go above and beyond from an ops perspective to get everything up and running.

And to be honest, that is okay. But the next step is up to the product team and the developers, to refactor or even rebuild parts of your solution to really leverage the power of a platform such as Azure Kubernetes Service.

It is the same thing we have seen in the past with lifting and shifting virtual machines to the public cloud. Basically, we switched *hosting provider* but are we really using the platform we selected?

Multi-cloud

Every software company wants a multi-cloud configuration, but it is difficult to achieve. However, it is a way to spread risk for your business as you are not locked into a single cloud, and sometimes will even save money. Public clouds provide services in different regions. Sometimes one region can be cheaper on one public cloud as opposed to another.

Then we also have customer requirements (remember, always think about the customers of the customer). Sometimes they will only allow one specific public cloud for compliance reasons. It is good to have a multi-cloud strategy, and at the very least, explore the options of running in multiple clouds.

This comes with a number of challenges. It is safe to say that if you are just using Kubernetes, then you should be good to go. But here's the thing: if we really want to use the platform and benefit from all its capabilities, then we are probably going to implement some cloud-specific features. For AKS, they would be features such as the Azure Container Networking Interface, Key Vault integration, and Azure Active Directory integration. Those are really specific and probably not available on something such as AWS.

An effective multi-cloud strategy doesn't limit itself to "just" using generic features. If you manage and implement your DevOps processes carefully, using multiple clouds and different (but similar) features per cloud is not a problem. It is just important to understand that multi-cloud doesn't mean you are simply "adding" another cloud to your portfolio; it requires both dev and ops investments.

Having said that, we do see more businesses researching a multi-cloud strategy. And even from Azure itself, solutions such as Azure Arc for Kubernetes can help you with that.

GitOps

GitOps is not new but its concept is growing more popular by the day. With GitOps we are basically integrating infrastructure as code, Git source control, and best practices from the world of software development. Picture this example: your Helm charts are living in a Git repository and your Kubernetes cluster monitors that repository. Once you merge the new additions in your Helm charts to the main branch of the repository, Kubernetes grabs the new Helm charts and applies them to the cluster.

That is one of the examples that you will see in the world of GitOps, and really the deployment itself is nothing special. But, integrating development best practices into your infrastructure as code and Kubernetes practices is something that you will really benefit from in the long run.

For AKS a popular GitOps solution is Flux (version 2), which can also be enabled from Azure itself. A popular alternative to Flux is Argo CD, which even comes with a fancy graphical user interface to manage your deployments.

Either way, GitOps is gaining popularity and is a logical step to take your deployments on Kubernetes to the next level. It is definitely worth investigating.

Security

Security is not necessarily *a future trend*. It is always relevant and will always be. However, with more businesses being digital and maturing, the focus is now shifting to security more and more. We see more and more businesses and community leaders diving into container security. Why? The same reason that we are now looking at cloud-native development. It's a result of the market maturing. Where initially the focus was on providing features and getting things up and running, the focus is now shifting more and more toward security. One could argue this should have been the case already, but this is just how it works.

We also see security being a priority for Microsoft, as we can see in the rapid development of new features for Microsoft Defender for Cloud and the additions for scanning container registries, running containers, and pretty much everything related to containers.

We see an increase in people implementing policies preventing containers from running privileged access and implementing guard rails, security boundaries, and proper logging and monitoring in general.

Where this was sometimes difficult and a lot of work in the past, there are now plenty of add-ons and third-party solutions for Kubernetes and AKS itself to quickly increase the level of your container and cluster security.

Azure Red Hat OpenShift

Azure Red Hat OpenShift is a collection of containerization technologies. Most of them are Kubernetes, but you will also find some Docker and additional runtimes in there. Azure Red Hat OpenShift is the result of a collaboration between Microsoft and Red Hat (the OpenShift platform) and is a platform-as-a-service offering for containers in Azure. Both Microsoft and Red Hat take responsibility for keeping your clusters up to date.

Red Hat has put a lot of work into OpenShift to include many of the best practices we have discussed during the course of this book as default with their offering. It does come with a price but it also comes with a lot of features.

Technically, you would still run Kubernetes but according to the Red Hat standard. Literally all Kubernetes management is abstracted away and operational management is done through OpenShift itself.

It is a new standard to "learn" and we would still recommend knowledge of Kubernetes in general as it is important to understand the moving parts your solution runs on.

One could say that OpenShift enforces a standard on you when it comes to security, networking, and monitoring. They are configured by default, which is a good thing. However, it really depends on your use case whether OpenShift is for you. The standard is good but it is a new platform to learn and you are dependent on what features Red Hat offers on its OpenShift platform. That and the fact that it is more expensive than AKS make most businesses reconsider using it. We do see the use case gaining more popularity but it is niche. If we come across it, it is usually at a large enterprise that already had a connection with Red Hat in one way or another.

A lot of these features are related to AKS. But what does the future of AKS look like? Let's take a look.

The future of AKS

As you may have gathered from reading this book, AKS is an ever-evolving service. Just in the last half of 2022, the networking part of AKS has grown from two **container network interfaces** (**CNIs**) to having the ability to bring your own CNI, the next iteration of the Azure CNI, with the added ability to dynamically assign IP addresses when needed. We then also got an extra feature of the Azure CNI, allowing it to use an overlay network. And near the end of the year, we got the option to run Azure CNI powered by Cilium, bringing an eBPF dataplane with it. Additionally, features such as Fleet allow us to manage AKS clusters at scale, and technologies such as Draft V2 allow us to quickly generate deployment files so we can also deploy at scale. In this section, we will take a look at each of these technologies.

eBPF

eBPF is a technology that allows sandboxed programs to be inserted into the Linux kernel. By doing this, you get enhanced traffic processing in the operating system's runtime. By using eBPF you get a rich set of use cases such as improved security, networking, observability, and even application tracing. So, this is definitely something to keep an eye on going forward.

Fleet

A trend we and Microsoft have been seeing over the last few years is customers wanting multiple small AKS clusters rather than one giant cluster that runs thousands of nodes. In fact, AKS currently supports up to 5,000 nodes in one cluster.

When it comes to managing multiple clusters, there are currently not many tools out there. Yes, we have GitOps to help deploy workloads to multiple clusters but that's just part of the story, as you may or may not know that looking after an AKS cluster is not the easiest of tasks. You have a lot of patching to do at both the OS level and the Kubernetes level. Say you have 100, or even 10 clusters to look after – that's not an easy task.

Luckily, Microsoft has a solution for this: **Kubernetes Fleet Manager**, or as we call it, **Fleet**. While it is still in early public preview, we see that it has promise and think it is definitely something to keep an eye on going forward.

Fleet is technically just another AKS cluster for you to manage, but this cluster is what is known as a hub cluster. Think hub-and-spoke networking here. This hub cluster is your main control plane. Here you can manage the upgrades of all your other clusters. It will have the ability for configuration policies to allow you to upgrade development clusters before production and push configuration to only development clusters and not **User Acceptance Testing** (**UAT**) or production by accident. Technically, Fleet allows us to isolate the environment into groups we are currently managing, but we can quickly change the group membership when we need to.

With Fleet, you even get the ability to load balance your North-South network traffic across workloads deployed in any number of fleet-managed clusters. This means you basically have a multi-region AKS cluster.

I am excited to see what Fleet has in store for us as it progresses out of preview. Hopefully, at some point, the hub cluster will be abstracted away from us and we will be able to manage all clusters via the Azure portal or infrastructure as code. If you wish to look further into Kubernetes Fleet Manager, check out the Azure docs: `https://learn.microsoft.com/azure/kubernetes-fleet/`

Draft v2

With all of the improvements to the operations side of AKS, you would understand it if there were no improvements to the development side, but there are. We now have a tool called **Draft v2**. This is the second version of **Draft**. But what actually is Draft? It is a tool that helps developers create not only Docker files from their source code, but also Kubernetes manifest files, Helm charts, kustomize configuration files, and even GitHub action workflows, all from the command line or via a VS Code extension.

Draft is the perfect tool for the developer who is just getting started in the world of containers. You run Draft in the directory that has your source code and it then scans your code to detect what language is being used. After it has detected the language, it will ask you some basic questions about your application and then create the needed files to build and deploy your container. It's worth noting that Draft is a good way to get started as it helps you containerize your solution, but for more complex apps you will have to make some changes to these files.

In the future, I am hoping that this tool will evolve and we will see some best practices and recommendations we can follow to improve the security and reliability of the deployed application. As this tool is open source, if you think you could help bring some helpful features, take a look at the Git repo: `https://github.com/Azure/draft/`.

Wasm

WebAssembly (**Wasm**) has been around for a while but at KubeCon North America 2022, it was probably the top topic alongside eBPF. It is a web standard developed by the W3C community group and is supported by the four main web browsers. Its binary format is designed for low latency and high performance to allow for near-native performance. You can see it as a technology built to be cloud native.

AKS is, at the time of writing, the best place to run Wasm in Azure. It uses containerd shims as its runtime (shims are essentially pieces of middleware to help translation from one technology to another). Containerd has the same runtime that the containers use in AKS. This means that your Wasm workloads will work over all your AKS node pools natively. So even if you have a Windows node pool and a Linux node pool, your Wasm workload will still work. This brings greater reliability to your workloads.

Wasm support for AKS is still in the early stages, but we feel this is where you will find a lot of development going on in the near future, so it's definitely one to watch. You can read more about AKS and Wasm at `https://learn.microsoft.com/azure/aks/use-wasi-node-pools`.

With all those technologies relevant to you now or in the future, it is important you have a thorough understanding of the platforms they run on. Certifications for Azure and Kubernetes specifically will help you achieve that understanding.

Azure and Kubernetes certifications

There are a number of certifications that can help you boost your knowledge specifically for Kubernetes. If you combine those certifications with Azure-related certifications, then you'll have a very good understanding of how you can leverage Azure Kubernetes Service.

Please note that even though these certifications are theoretical and performance-based, real-world experience is key to successfully operating Azure Kubernetes Service. However, gaining that real-world experience is much easier if you have a solid base, such as certifications.

Let's take a look at what certifications are relevant for your AKS journey.

AZ-305: Designing Microsoft Azure Infrastructure Solutions

Designing Microsoft Azure Infrastructure Solutions is often referred to as the *architecture* certification. That may sound like a lot and you might be thinking, *"But I just want to do Kubernetes."* As you have learned throughout this book, AKS integrates with many different Azure technologies. From Key Vault to virtual networks to storage, there are integration capabilities everywhere. Understanding the platform that your solution runs on will help you think about and select the right integrations that will benefit your solution.

Even if you are a developer, this certification will be of value to you.

> **Important note**
>
> If you are aiming to obtain the Microsoft Certified: Azure Solutions Architect Expert credentials, then the AZ-104 certification exam is a prerequisite. This certification is also very good preparation for the AZ-305 exam.

You can find more information about the learning path, the exam outline, and much more here: `https://learn.microsoft.com/en-us/certifications/exams/az-305`.

Certified Kubernetes Application Developer

In the *Future trends* section of this chapter, we talked about cloud-native development. Development starts with the selection and knowledge of the platform. In this case, that platform is Kubernetes. Modernizing your solution once you have decided to go with container technologies and Kubernetes is a whole lot easier if you have learned about the platform.

By studying for **Certified Kubernetes Application Developer** (**CKAD**), you will learn about best practices when creating container images and how to maintain them, run them, and debug them. Now, we would go as far as to say that this certification is not just for developers. When managing and maintaining a Kubernetes environment, there is a big gray area in responsibilities between developers and operations. For some teams, operations are limited to keeping the cluster up and running while the developers take responsibility for writing the configuration that runs on the cluster (such as the configmaps, deployments, etc.). In other teams, the developers focus on just providing a container image and requesting features and configuration from the operations team.

There is no right or wrong answer here, but anyone in the operations team will benefit from this certification.

The Cloud Native Computing Foundation maintains a curriculum that outlines the requirements of the certification and thus the things you will learn during the preparation for CKAD. The curriculum can be found here: `https://github.com/cncf/curriculum/blob/master/CKAD_Curriculum_v1.26.pdf`.

Certified Kubernetes Administrator

Certified Kubernetes Administrator (**CKA**) is focused on the operations side of things. This certification goes far beyond what you need to know to deploy Azure Kubernetes Service as it is focused on managing and troubleshooting regular Kubernetes clusters. But that is a good thing. We come across plenty of teams who manage and maintain AKS environments but without in-depth knowledge of the moving parts of Kubernetes itself. You don't need to manage the etcd database or configure the scheduler on AKS as Microsoft does that for you. However, it is vital to understand how they work when troubleshooting any Kubernetes cluster in any public cloud.

In fact, after achieving this certification, you will be able to deploy your own vanilla Kubernetes cluster, anywhere. Why would you do that? Well, earlier we mentioned hybrid and multi-cloud setups are becoming more in demand. Being able to deploy and run Kubernetes anywhere is a very useful skill to have!

You can find the outline for this certification here: `https://github.com/cncf/curriculum/blob/master/CKA_Curriculum_v1.26.pdf`.

Certified Kubernetes Security Specialist

The **Certified Kubernetes Security Specialist** (**CKS**) certification is something that we see as a good addition to both CKA and CKAD. Security knowledge is something that should not be limited to security specialists. Everyone, whether they work in operations or development, should have a good understanding of the security capabilities the platform provides.

That doesn't mean that developers and operations should manage all the aspects of the security posture of the platform on a daily basis, but an understanding of it is crucial. We don't want a security team to enforce security policies that might block a feature of the application and end up breaking it. On the other hand, we also don't want deployments and solutions to be insecure by default. Security is a shared responsibility.

The CKS certification is focused on hardening your cluster, network security, supply chain security, and runtime security and monitoring. You could call these hygiene factors. The reason why we think this certification is so valuable is that often we start with building a cluster, eventually going to production, but never really take the time to stop and think about security – not until a customer asks or management decides it is time for the annual security tests. CKS will help you understand security mechanisms and implement them successfully on any Kubernetes cluster – definitely worth it!

You can find the curriculum for CKS here: `https://github.com/cncf/curriculum/blob/master/CKS_Curriculum_%20v1.26.pdf`.

Linux Foundation Certified System Administrator

The majority of containers run on Linux; in fact, Kubernetes itself runs on Linux. Troubleshooting when a container cannot run successfully is specific to Kubernetes but, usually, it doesn't stop there. You want to take a look inside the containers. Debugging containers on Kubernetes is a very useful skill to have. A scenario we never want is where the operations team responds to the developers with "*The Pods are running! Developers, good luck, we have done our job,*" when, in fact, the operations team probably understands the Kubernetes platform itself better than most developers. You (the operations team) understand what impact the cluster configuration can have on the configuration inside the container as well. It is your responsibility to relay that information to your development colleagues. For example, are all environment variables passed by your configmap and does the application inside the container actually see them?

Being able to debug configuration issues inside the container requires Linux knowledge as most containers run on Linux-based images. Having the knowledge that comes with the **Linux Foundation Certified System Administrator** (LFCS) certification will make you a valuable member of the team and help you bridge the gap between operations and development.

The LFCS certification is offered by the Linux Foundation and more information can be found here: `https://training.linuxfoundation.org/certification/linux-foundation-certified-sysadmin-lfcs/`.

Summary

In this chapter, we have learned what additions to Kubernetes there will be in the future. Yes, there will be more initiatives by the time this book is released – even during the writing of this chapter, people were working on new initiatives in the world of cloud-native computing.

We have looked at upcoming technologies that are starting to gain ground in the world of Kubernetes. Technologies such as Fleet and Draft v2 make life easier for operations teams, and new features such as Wasm are game-changing for developers.

Finally, we have learned about valuable certifications that will help you kick-start your AKS journey. They're not just useful because it's good to have a certification, but because the road toward those certifications will give you the knowledge to successfully operate container technologies on the Microsoft Azure platform.

That brings us to the final paragraph of this book. The world of containers is enormous and evolving rapidly. By the time you have finished reading this, new initiatives will have launched, but we are confident this book will give you a solid basis for getting started in the world of containers on Azure. The best advice we can give you is to keep learning and keep experimenting.

Index

A

B

C

D

M

N

O

P

R

S

W

Other Books You May Enjoy

If you enjoyed this book, you may be interested in these other books by Packt:

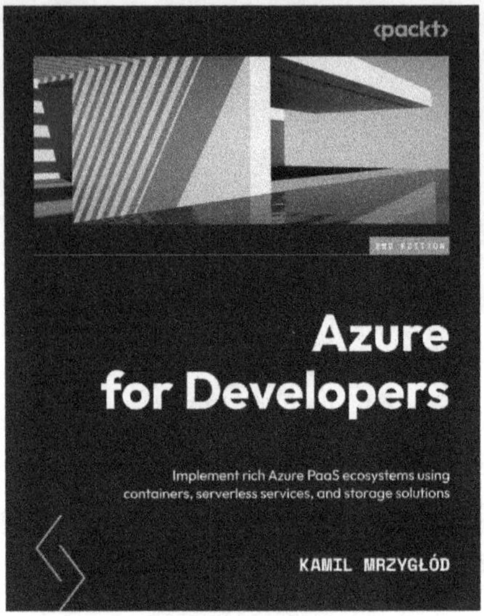
‹packt›
Azure
for Developers
Implement rich Azure PaaS ecosystems using
containers, serverless services, and storage solutions
KAMIL MRZYGŁÓD

Azure for Developers - Second Edition

Kamil Mrzygłód

ISBN: 978-1-80324-009-1

- Identify the Azure services that can help you get the results you need
- Implement PaaS components – Azure App Service, Azure SQL, Traffic Manager, CDN, Notification Hubs, and Azure Cognitive Search
- Work with serverless components
- Integrate applications with storage
- Put together messaging components (Event Hubs, Service Bus, and Azure Queue Storage)
- Use Application Insights to create complete monitoring solutions
- Secure solutions using Azure RBAC and manage identities
- Develop fast and scalable cloud applications

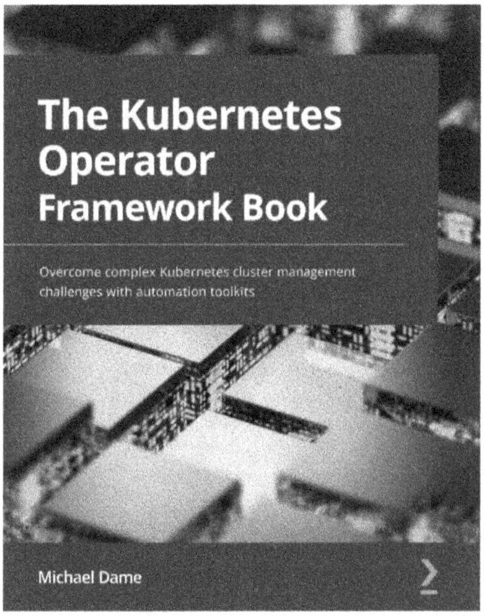
The Kubernetes Operator Framework Book
Overcome complex Kubernetes cluster management challenges with automation toolkits
Michael Dame

The Kubernetes Operator Framework Book

Michael Dame

ISBN: 978-1-80323-285-0

- Gain insight into the Operator Framework and the benefits of operators
- Implement standard approaches for designing an operator
- Develop an operator in a stepwise manner using the Operator SDK
- Publish operators using distribution options such as OperatorHub.io
- Deploy operators using different Operator Lifecycle Manager options
- Discover how Kubernetes development standards relate to operators
- Apply knowledge learned from the case studies of real-world operators

Packt is searching for authors like you

If you're interested in becoming an author for Packt, please visit `authors.packtpub.com` and apply today. We have worked with thousands of developers and tech professionals, just like you, to help them share their insight with the global tech community. You can make a general application, apply for a specific hot topic that we are recruiting an author for, or submit your own idea.

Share Your Thoughts

Now you've finished *Azure Containers Explained*, we'd love to hear your thoughts! Scan the QR code below to go straight to the Amazon review page for this book and share your feedback or leave a review on the site that you purchased it from.

`https://packt.link/r/180323105X`

Your review is important to us and the tech community and will help us make sure we're delivering excellent quality content.

Download a free PDF copy of this book

Thanks for purchasing this book!

Do you like to read on the go but are unable to carry your print books everywhere? Is your eBook purchase not compatible with the device of your choice?

Don't worry, now with every Packt book you get a DRM-free PDF version of that book at no cost.

Read anywhere, any place, on any device. Search, copy, and paste code from your favorite technical books directly into your application.

The perks don't stop there, you can get exclusive access to discounts, newsletters, and great free content in your inbox daily

Follow these simple steps to get the benefits:

1. Scan the QR code or visit the link below

   `https://packt.link/free-ebook/9781803231051`

2. Submit your proof of purchase
3. That's it! We'll send your free PDF and other benefits to your email directly

www.ingramcontent.com/pod-product-compliance
Lightning Source LLC
Chambersburg PA
CBHW060528060326
40690CB00017B/3416
* 9 7 8 1 8 0 3 2 3 1 0 5 1 *